Catalysis by Metals on Perovskite-Type Oxides

Catalysis by Metals on Perovskite-Type Oxides

Editor
Davide Ferri

MDPI • Basel • Beijing • Wuhan • Barcelona • Belgrade • Manchester • Tokyo • Cluj • Tianjin

Editor
Davide Ferri
Paul Scherrer Institut
Switzerland

Editorial Office
MDPI
St. Alban-Anlage 66
4052 Basel, Switzerland

This is a reprint of articles from the Special Issue published online in the open access journal *Catalysts* (ISSN 2073-4344) (available at: https://www.mdpi.com/journal/catalysts/special_issues/Perovskite).

For citation purposes, cite each article independently as indicated on the article page online and as indicated below:

LastName, A.A.; LastName, B.B.; LastName, C.C. Article Title. *Journal Name* **Year**, *Volume Number*, Page Range.

ISBN 978-3-03943-697-2 (Hbk)
ISBN 978-3-03943-698-9 (PDF)

© 2021 by the authors. Articles in this book are Open Access and distributed under the Creative Commons Attribution (CC BY) license, which allows users to download, copy and build upon published articles, as long as the author and publisher are properly credited, which ensures maximum dissemination and a wider impact of our publications.

The book as a whole is distributed by MDPI under the terms and conditions of the Creative Commons license CC BY-NC-ND.

Contents

About the Editor vii

Davide Ferri
Catalysis by Metals on Perovskite-Type Oxides
Reprinted from: *Catalysts* **2020**, *10*, 1062, doi:10.3390/catal10091062 1

María Luisa Rojas-Cervantes and Eva Castillejos
Perovskites as Catalysts in Advanced Oxidation Processes for Wastewater Treatment
Reprinted from: *Catalysts* **2019**, *9*, 230, doi:10.3390/catal9030230 5

Bae-Jung Kim, Emiliana Fabbri, Ivano E. Castelli, Mario Borlaf, Thomas Graule, Maarten Nachtegaal and Thomas J. Schmidt
Fe-Doping in Double Perovskite PrBaCo$_{2(1-x)}$Fe$_{2x}$O$_{6-\delta}$: Insights into Structural and Electronic Effects to Enhance Oxygen Evolution Catalyst Stability
Reprinted from: *Catalysts* **2019**, *9*, 263, doi:10.3390/catal9030263 43

Shaoxia Guo, Guilong Liu, Tong Han, Ziyang Zhang and Yuan Liu
K-Modulated Co Nanoparticles Trapped in La-Ga-O as Superior Catalysts for Higher Alcohols Synthesis from Syngas
Reprinted from: *Catalysts* **2019**, *9*, 218, doi:10.3390/catal9030218 61

Patrick Steiger, Dariusz Burnat, Oliver Kröcher, Andre Heel and Davide Ferri
Segregation of Nickel/Iron Bimetallic Particles from Lanthanum Doped Strontium Titanates to Improve Sulfur Stability of Solid Oxide Fuel Cell Anodes
Reprinted from: *Catalysts* **2019**, *9*, 332, doi:10.3390/catal9040332 75

Mohammed Ismael and Michael Wark
Perovskite-type LaFeO$_3$: Photoelectrochemical Properties and Photocatalytic Degradation of Organic Pollutants Under Visible Light Irradiation
Reprinted from: *Catalysts* **2019**, *9*, 342, doi:10.3390/catal9040342 91

Antonella Glisenti and Andrea Vittadini
On the Effects of Doping on the Catalytic Performance of (La,Sr)CoO$_3$. A DFT Study of CO Oxidation
Reprinted from: *Catalysts* **2019**, *9*, 312, doi:10.3390/catal9040312 107

Craig Aldridge, Verónica Torregrosa-Rivero, Vicente Albaladejo-Fuentes, María-Salvadora Sánchez-Adsuar and María-José Illán-Gómez
BaTi$_{0.8}$B$_{0.2}$O$_3$ (B = Mn, Fe, Co, Cu) LNT Catalysts: Effect of Partial Ti Substitution on NOx Storage Capacity
Reprinted from: *Catalysts* **2019**, *9*, 365, doi:10.3390/catal9040365 115

Ata ul Rauf Salman, Signe Marit Hyrve, Samuel Konrad Regli, Muhammad Zubair, Bjørn Christian Enger, Rune Lødeng, David Waller and Magnus Rønning
Catalytic Oxidation of NO over LaCo$_{1-x}$B$_x$O$_3$ (B = Mn, Ni) Perovskites for Nitric Acid Production
Reprinted from: *Catalysts* **2019**, *9*, 429, doi:10.3390/catal9050429 127

Bertrand Heidinger, Sébastien Royer, Houshang Alamdari, Jean-Marc Giraudon and Jean-François Lamonier
Reactive Grinding Synthesis of LaBO$_3$ (B: Mn, Fe) Perovskite; Properties for Toluene Total Oxidation
Reprinted from: *Catalysts* **2019**, *9*, 633, doi:10.3390/catal9080633 . . . . **139**

About the Editor

Davide Ferri studied industrial chemistry at the University of Milano (Italy) and received his PhD in chemistry at the ETH Zurich in 2002. After a short experience at Bruker Optics GmbH (Switzerland), he was assistant professor at ETH Zurich, and then group head at the Swiss Federal Laboratories for Material Science and Technology (Empa). He is currently head of the group Applied Catalysis and Spectroscopy at the Paul Scherrer Institut (Switzerland). His interests span from environmental catalysts to liquid phase catalyzed reactions and the development of in situ/operando experiments. He uses vibrational spectroscopies in combination with X-ray-based methods.

Editorial

Catalysis by Metals on Perovskite-Type Oxides

Davide Ferri

Paul Scherrer Institut, Forschungsstrasse 111, CH-5232 Villigen, Switzerland; davide.ferri@psi.ch

Received: 9 September 2020; Accepted: 11 September 2020; Published: 15 September 2020

Perovskites are currently on everyone's lips and have made it in high-impact scientific journals because of the revolutionary hybrid organic–inorganic lead halide perovskite materials for solar cells. The mixed metal oxide counterparts have received equal interest in a variety of technological applications for decades. The interest for perovskite-type oxides lies in their vast compositional and structural variability that can be exploited to tailor physico-chemical properties such as oxygen mobility and vacancies, redox, as well as electronic and ionic conductivities for specific technical applications. Besides oxygen mobility and the variety of element combinations adopting the perovskite-type structure, a further property is interesting for catalytic applications—the possibility to exploit them as a precursor of active catalysts upon exsolution of transition metals in the form of particles at their surface. Transition metals have been shown to experience reversible segregation: Reduced metal nanoparticles are exposed at the moment they need to be used for a catalytic process. The reversible segregation allows their protection from poisoning and growth, thus regenerating the original mixed oxide structure. The interaction between exsolved particles and defective mixed metal oxides also results in potentially new material properties that are difficult to obtain by material synthesis or by combination of the same transition element and other metal oxide supports.

In this Special Issue, various aspects of the synthesis, structure, as well as sorptive and catalytic properties of perovskite-type mixed oxides are presented for a wide range of catalytic processes, demonstrating once more the versatility of this class of materials.

Rojas-Cervantes and co-workers review the utility of perovskite-type oxide-based oxidation catalysts for the treatment of wastewater using various oxidants, as well as ultraviolet-visible irradiation to promote photocatalysis [1]. The authors analyze physico-chemical properties of perovskite-type oxides such as mobility of lattice oxygen and oxygen vacancy formation, stabilizing unusual oxidation states of the contributing elements that can be exploited to enhance advanced oxidation processes (AOP), making use of radicals.

In the contribution of the $PrBaCo_{2(1-x)}Fe_{2x}O_{6-\delta}$ layered double perovskite used as an electrocatalyst for the oxygen evolution reaction (OER) in alkaline media, Kim et al. [2] demonstrate the promoting effect of Fe using X-ray absorption spectroscopy. Fe introduction into $PrBaCo_2O_{6-\delta}$ forces Co to adopt a lower oxidation state to allow for charge compensation. This stabilizes Co and the layered double perovskite against dissolution under the reaction conditions but sustains the formation of the active surface Co oxyhydroxide layer.

Guo et al. [3] demonstrate the feasibility of using $LaCo_yGa_{1-y}O_3$ mixed oxides to catalyze the synthesis of alcohols (methanol/ethanol predominantly) from syngas. $La_{1-x}K_xCo_{0.65}Ga_{0.35}O_3$ is used as the precursor of segregated Co nanoparticles that are embedded within the La-Ga-O composite oxide, thus improving their stability in the reaction atmosphere. The role of K is found to increase the atomic dispersion of Co and improve the coking resistance of the composite catalyst by formation of La_2O_3.

In the work by Steiger et al. [4], the reversible segregation of Ni as an active element for the water gas shift reaction, and a second transition metal is studied to explore the sulfur tolerance in solid oxide fuel cells. Only Fe out of Mn, Mo, Cr, and Fe is found to increase the sulfur tolerance of $La_{0.3}Sr_{0.55}Ti_{0.95}Ni_{0.05}O_{3\pm\delta}$. The simultaneous segregation of Fe and Ni at high temperature does not hamper the reversible segregation–reintegration of the two metals within the perovskite mixed oxide upon oxidation, allowing for long-term use at high temperature.

Wark and co-workers present the photocatalytic activity of $LaFeO_3$ toward Rhodamine B decomposition and screen the optimal conditions to obtain the highest performance [5]. Besides textural characterization, photoelectrochemical characterization was exploited to explain the different behavior with calcination temperature, showing that lower temperatures promoted the photo-induced charge carrier transfer and separation efficiency.

A density functional theory study by Glisenti et al. [6] shows that substitution of La by Sr at the A-site of $LaCoO_3$ lowers the formation energy of oxygen vacancies that is beneficial for CO oxidation in three-way catalysts. Similar effects can be obtained at the B-site only by substitution of Co by Cu. Substitution effects appear to be greater in $SrTiO_3$.

The effect of partial substitution of Ti in $BaTiO_3$ used as a lean NO_x trap catalyst is studied by Aldridge et al. [7] with respect to the influence on NO_x storage capacity. Especially, the use of Cu is beneficial as it promotes the segregation of Ba_2TiO_4 and NO_x storage. $BaTi_{0.8}Cu_{0.2}O_3$ exhibits the highest amount of oxygen vacancies and a storage capacity in the range of highly active noble metal-based catalysts.

A series of $LaCo_{1-x}Mn_xO_3$ and $LaCo_{1-y}Ni_yO_3$ catalysts are tested by Rønning and coworkers for NO oxidation to NO_2 that is a crucial step in the production of nitric acid [8]. While $LaCoO_3$ exhibits the highest activity amongst the undoped perovskites, $LaCo_{0.75}Ni_{0.25}O_3$ and $LaCo_{0.75}Mn_{0.25}O_3$ are found to be optimum in substituted catalysts, showing that perovskites are promising catalysts for NO oxidation in industrial conditions.

Finally, a three-step reactive grinding process including solid-state synthesis, high-energy ball milling, and low-energy ball milling in wet conditions followed by calcination at 400 °C is used to prepare $LaMnO_3$ and $LaFeO_3$ by Heidinger et al. [9]. In both cases, the catalytic performance for toluene oxidation increases after each synthesis step of the process in line with the increase in specific surface area that reaches ca. 19 $m^2 \cdot g^{-1}$ for $LaFeO_3$.

I take the opportunity to thank the principal investigators and all authors of the contributed publications for their effort and for willing to make this Special Issue on perovskite-type oxides possible.

Funding: The Swiss National Science Foundation (SNF) is acknowledged for funding (project nr. 200021_159568).

Conflicts of Interest: The author declares no conflict of interest.

References

1. Rojas-Cervantes, M.L.; Castillejos-Lopéz, E. Perovskites as Catalysts in Advanced Oxidation Processes for Wastewater Treatment. *Catalyst* **2019**, *9*, 230. [CrossRef]
2. Kim, B.J.; Fabbri, E.; Castelli, I.E.; Borlaf, M.; Graule, T.; Nachtegaal, M.; Schmidt, T.J. Fe-Doping in Double Perovskite $PrBaCo_{2(1-x)}Fe_{2x}O_{6-\delta}$: Insights into Structural and Electronic Effects to Enhance Oxygen Evolution Catalyst Stability. *Catalyst* **2019**, *9*, 263. [CrossRef]
3. Guo, S.; Liu, G.; Han, T.; Zhang, Z.; Liu, Y. K-Modulated Co Nanoparticles Trapped in La-Ga-O as Superior Catalysts for Higher Alcohols Synthesis from Syngas. *Catalyst* **2019**, *9*, 218. [CrossRef]
4. Steiger, P.; Burnat, D.; Kröcher, O.; Heel, A.; Ferri, D. Segregation of Nickel/Iron Bimetallic Particles from Lanthanum Doped Strontium Titanates to Improve Sulfur Stability of Solid Oxide Fuel Cell Anodes. *Catalyst* **2019**, *9*, 332. [CrossRef]
5. Ismael, M.; Wark, M. Perovskite-type $LaFeO_3$: Photoelectrochemical Properties and Photocatalytic Degradation of Organic Pollutants under Visible Light Irradiation. *Catalyst* **2019**, *9*, 342. [CrossRef]
6. Glisenti, A.; Vittadini, A. On the Effects of Doping on the Catalytic Performance of (La,Sr)CoO_3. A DFT Study of CO Oxidation. *Catalyst* **2019**, *9*, 312. [CrossRef]
7. Aldridge, C.; Torregrosa-Rivero, V.; Albaladejo-Fuentes, V.; Sanchez-Adsuar, M.S.; Illan-Gomez, M.J. $BaTi_{0.8}B_{0.2}O_3$ (B = Mn, Fe, Co, Cu) LNT catalysts: Efect of partial ti substitution on nox storage capacity. *Catalysts* **2019**, *9*, 365. [CrossRef]

8. Salman, A.R.; Hyrve, S.M.; Regli, S.K.; Zubair, M.; Enger, B.C.; Lodeng, R.; Waller, D.; Ronning, M. Catalytic oxidation of NO over LaCo$_{1-x}$B$_x$O$_3$ (B= Mn, Ni) perovskites for nictric acid production. *Catalysts* **2019**, *9*, 429. [CrossRef]
9. Heidinger, B.; Royer, S.; Alamdari, H.; Giraudon, J.M.; Lamonier, J.F. Reactive Grinding Synthesis of LaBO$_3$ (B: Mn, Fe) Perovskite; Properties for Toluene Total Oxidation. *Catalysts* **2019**, *9*, 633. [CrossRef]

© 2020 by the author. Licensee MDPI, Basel, Switzerland. This article is an open access article distributed under the terms and conditions of the Creative Commons Attribution (CC BY) license (http://creativecommons.org/licenses/by/4.0/).

Review

Perovskites as Catalysts in Advanced Oxidation Processes for Wastewater Treatment

María Luisa Rojas-Cervantes * and Eva Castillejos

Departamento de Química Inorgánica y Química Técnica, Facultad de Ciencias, UNED. Paseo Senda del Rey nº 9, Madrid 28040, Spain; castillejoseva@ccia.uned.es
* Correspondence: mrojas@ccia.uned.es

Received: 1 February 2019; Accepted: 13 February 2019; Published: 2 March 2019

Abstract: Advanced oxidation processes (AOPs), based on the formation of highly reactive radicals are able to degrade many organic contaminants present in effluent water. In the heterogeneous AOPS the presence of a solid which acts as catalyst in combination with other systems (O_3, H_2O_2, light) is required. Among the different materials that can catalyse these processes, perovskites are found to be very promising, because they are highly stable and exhibit a high mobility of network oxygen with the possibility of forming vacancies and to stabilize unusual oxidation states of metals. In this review, we show the fundamentals of different kinds of AOPs and the application of perovskite type oxides in them, classified attending to the oxidant used, ozone, H_2O_2 or peroxymonosulfate, alone or in combination with other systems. The photocatalytic oxidation, consisting in the activation of the perovskite by irradiation with ultraviolet or visible light is also revised.

Keywords: perovskites; advanced oxidation processes (AOPs); Fenton-like; peroxymonosulfate; heterogeneous photocatalysis

1. Introduction

Advanced oxidation processes (AOPs) are based on the generation of radical intermediates, mainly hydroxyl radicals (HO•), in amount enough to be able to attack and oxidize either partially or fully most of the recalcitrant chemicals present in the effluent water, such as pesticides, dyes, pharmaceuticals and so on [1]. The processes based on the free radicals occur at higher rates of degradation than those based on other chemical oxidation technologies and are not highly selective [2,3]. The high oxidation potential of hydroxyl radicals (2.80 v) make them capable of attacking organic compounds by abstracting a hydrogen atom or by adding to the double bonds, carrying out their mineralization by transformation into more oxidized intermediates, carbon dioxide, water and inorganic salts. These reactions of hydroxyl radicals with organic compounds can be written as follows:

$$HO• + R-H \rightarrow H_2O + R• \tag{1}$$

$$HO• + C=C \rightarrow HO-C-C• \tag{2}$$

$$HO• + Ph-H \rightarrow Ph-H(OH)• \tag{3}$$

AOPs can be classified in several categories, depending on the different reagent systems used for the generation of hydroxyl radicals. Attending to the reaction medium, these advanced oxidation processes can be classified either as homogeneous or heterogeneous [4]. The first ones can be subdivided in turn in those using energy (ultraviolet or visible radiation, ultrasound energy, electrical energy) and those not involving energy (ozone (O_3) in alkaline medium, O_3/H_2O_2 and H_2O_2/homogeneous catalyst, generally Fe^{2+}, known as Fenton process). The heterogeneous processes can be classified in four main groups: (i) catalytic ozonation, which uses the combination of O_3 and a solid catalyst; (ii) photocatalytic ozonation, under the action of O_3/light (UV or visible)/solid catalyst; (iii) Fenton-like processes, which are produced by the action of H_2O_2/solid catalyst, containing mainly the Fe^{2+}/Fe^{3+} couple but also other transition metal ions with multiple oxidation states; when they are combined with the action of light they are called Photo-Fenton processes; and iv) photocatalytic oxidation, by combination of light (UV or visible) and a solid catalyst.

It must be remarked that the single ozonation or the use of only H_2O_2 belong to the class of chemical oxidation technologies, as they work on the direct attack of the oxidants, not being considered as AOPS, because they do no generate hydroxyl radicals by themselves [5]. Only when O_3 and H_2O_2 are combined between them or its individual action is supplemented by other dissipating energy components, such as UV/visible light or ultrasound or by activation with a catalyst, the formation of free radicals occurs and they can be considered AOPs.

In the last fifteen years some reviews in literature have devoted to show the state of art of AOPS for wastewater treatment [4–7] and in particular Fenton and photo-Fenton processes [8,9]. Most of these oxidation technologies are usually expensive and in addition, they are unable to completely degrade the organic compounds present in real wastewater and they cannot process the large volumes of waste generated. However, AOPs can degrade the residue up to a certain level of toxicity and then the intermediate can be furthered degraded by the conventional methods. Furthermore, the combination of AOPs (as a pre-treatment or post-treatment stage) with a biological treatment contributes to reduce operating costs of the global process [10].

As mentioned above, the heterogeneous advanced oxidation processes generally use solid catalysts in combination with other systems (O_3, H_2O_2, light) to carry out the degradation of organics. The main advantage of heterogeneous catalysts with respect to the homogeneous ones is the facility of separation of the product and of the recovery of the catalyst. However, to be applied in the industry, heterogeneous catalysts must satisfy some specifications, such as high activity, thermal, mechanical, physical and chemical stability and resistance to the deactivation.

Perovskite-type oxides of the general formula ABO_3, where A is a rare earth metal and B a transition metal, have attracted the attention of many scientists because of their unique structural features. They have a well-defined structure, which allows the introduction of a wide variety of metal ions in both A and B positions [11,12]. Its structure is represented in Figure 1. The partial substitution of these cations by other foreign leads to changes in the oxidation states of metal ions and to the formation of oxygen vacancies. The thermal and hydrothermal stability of perovskites is quite high and as a result, they can be applied to gas or solid reactions carried out at high temperatures or liquid-phase reactions occurring at low temperatures [13,14]. The high mobility of network oxygen and the stabilization of unusual oxidation states confer them a diversity or properties, which allows their application as solid oxide fuel cells [15], magnetic and electrode materials [16], chemical sensors [17], adsorbents [18] and heterogeneous catalysts in industrial reactions [11,14,19]. One of the potential applications of perovskites is as catalysts in carbon-based electrodes, which has incited many scientists to study the mechanism of the catalytic decomposition of H_2O_2 by perovskites [20–22].

Catalysts used in oxidation technologies can be classified as follows: (i) metal catalysts, usually supported on a metal oxide surface (TiO_2, Al_2O_3, ZrO_2 and CeO_2) or on active carbon; (ii) metal oxide catalysts and (iii) organometallic catalysts. Different heterogeneous catalysts have been applied in some AOPs. As an example, the use of several heterogeneous systems containing iron species stabilized in a host matrix, such as oxides [23–26], clays [23,27–30], zeolites [29,31,32] or carbon materials [33–37] in the Fenton and the Fenton-like processes has been reported. In the last years perovskites have been applied as Fenton catalysts because of their versatile composition and high stability [19]. Furthermore, the existence of redox active sites in B cations and oxygen vacancies may facilitate the transformation of H_2O_2 into HO• [38–40].

There are some reviews in literature describing the employment of clays and mineral oxides, mainly of iron, in Fenton-like processes [23,25,26]. However, until our knowledge, there is no any review reporting the use of perovskites in AOPs. In the present paper we describe the utilization of perovskites and like-perovskite oxides in this kind of processes, classified as a function of the oxidant reactant responsible for generating the free radicals, by alone or combined with other systems, giving place to the hybrid methods. Thus, in first place, in Section 2, we report the processes based on ozone, O_3, including catalytic ozonation (O_3/catalyst) and photocatalytic ozonation (O_3/catalyst/light). In the following section we revise the use of H_2O_2 as oxidant in three kinds or systems: (i) Fenton-like reactions (H_2O_2/catalyst); (ii) photo Fenton-like reactions (H_2O_2/catalyst/light); and (iii) catalytic wet peroxide oxidation, CWPO, (H_2O_2/catalyst/air). In the last fifteen years a new chemical oxidant, peroxymonosulfate (PMS), has aroused a great interest as alternative to others as H_2O_2 or O_3. For that reason, the fourth section is devoted to describing the use of PMS, activated by a catalyst (perovskite) or by the combination of both a catalyst and light irradiation. The photocatalytic oxidation, consisting in the activation of perovskite by irradiation with UV or visible light, is revised in Section 5. Finally, a short section devoted to the degradation under dark ambient conditions, show some examples in which perovskites catalyse the oxidation of organics in the absence of light and without additional chemical oxidant. Although these processes should not be considered strictly AOPS, because radicals are not formed, some authors consider them as novel advanced oxidation technologies for low cost treatment of wastewaters.

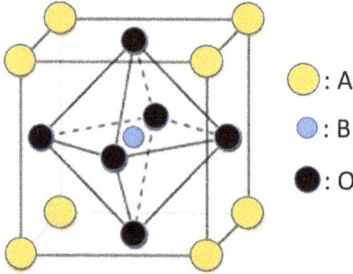

Figure 1. Unit cell of perovskite centred on A.

2. Processes Based on Ozone

Heterogeneous catalytic ozonation is getting an enormous attention in the treatment of drinking and waste water, because of its capacity to improve the mineralization and degradation of organic pollutants, its manufacturing simplicity and economic nature. There are many advantages to the use of ozone compared to other conventional technologies due to its high oxidizing power (2.07 V). Ozone can degrade organic pollutants by direct electrophilic attack with molecular ozone or by indirect attack with hydroxyl radicals HO•, which are generated through its decomposition process. Single ozonation has been widely used in water and wastewater because is an effective oxidation process. However, several disadvantages can limit its application such as slow and incomplete oxidation or mineralization (the intermediates are not totally oxidised to CO_2 and water) and the low solubility and stability of ozone in water. At first sight, heterogeneous catalytic ozonation does not present these drawbacks although much research is still needed. Catalytic ozonation utilizes solid catalysts in order to improve the decomposition of ozone and to enhance the production of hydroxyl radicals, HO•. In general, the heterogeneous catalysts decompose the ozone into caged or free radicals or simply they adsorb reactants facilitating their reaction. To date, several types of heterogeneous ozonation catalysts such as metal oxides, supported metal oxides and carbon materials have been tested with promising results. Perovskites-type metal oxides constitute undoubtedly an interesting alternative due to the high stability under aggressive conditions, high degree of stabilization of transition metals in their oxidation states and high oxygen mobility. However, the catalytic ozonation mechanism with perovskites is still a challenge for chemists and not many detailed studies are available. Most of the studies have been carried out by the same research group [41–44].

The first studies concerning catalytic ozonation with active perovskites appeared in 2006 [41]. The authors studied the ozonation decomposition of pyruvic acid, a refractory substance typically produced after oxidation of phenol-like compounds, in the presence of $LaTi_{0.15}Cu_{0.85}O_3$ perovskite. This perovskite was active and stable in the ozonation process being oxalic and acetic acids the only intermediates formed. Experimental results clearly indicated that typical operating parameters like ozone concentration, mass of catalyst or temperature, performed a key role on the pyruvic acid ozonation. The catalyst exhibited high stability and its catalytic activity improved after the first use. For instance, after 150 min of reaction the pyruvic acid had practically been eliminated, in contrast to the 67% of conversion achieved with fresh catalyst for the same period. Regarding kinetic considerations, authors proposed a Langmuir–Hinshelwood mechanism derived from bi-adsorption of pyruvic acid and ozone on different active sites and successive reaction of pyruvic acid with the O• radicals on the surface through reactions 1 and 2, before occurring the desorption of formed products (see Scheme 1).

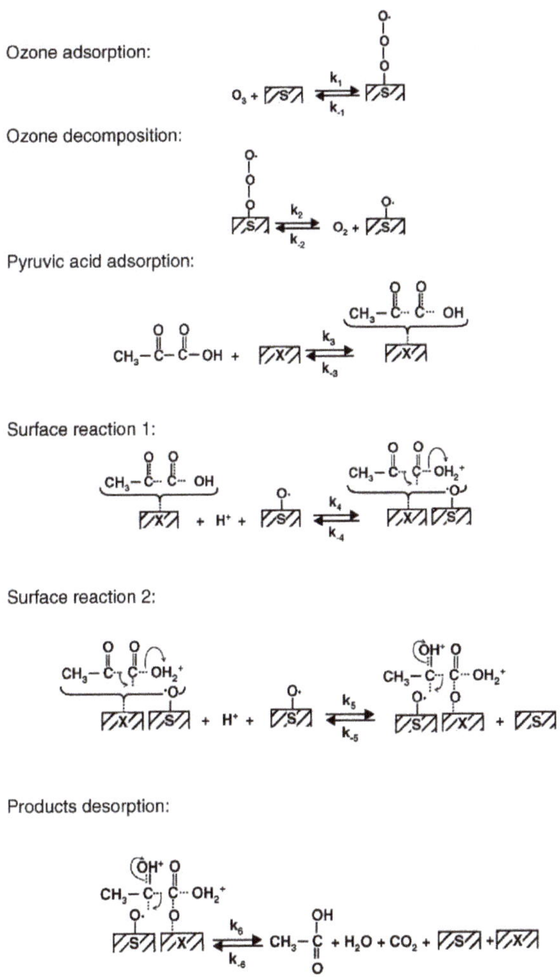

Scheme 1. Mechanism of catalytic ozonation of piruvic acid in the presence of $LaTi_{0.15}Cu_{0.85}O_3$. With permission from [41].

Carbajo et al. [42] extended the study of the removal of pyruvic acid from water through catalytic ozonation to other perovskites, $LaTi_{1-x}Cu_xO_3$ and $LaTi_{1-x}Co_xO_3$, which were compared with other catalysts, such as $Ru-Al_2O_3$, $Ru-CeO_2$, $FeO(OH)$ and MCM-41, this last impregnated with copper or cobalt. The results showed that only perovskites and $Ru-CeO_2$ catalysts increased significantly the pyruvic acid depletion with respect to the produced in the absence of catalyst. Conversion values of 80% were reached after 2 h of catalytic ozonation. Due to the low adsorption capacity of the catalysts to adsorb the pyruvic molecules, authors concluded that the catalytic ozonation mechanism was governed by surface reactions involving adsorbed ozone and dissolved pyruvic acid.

Another pioneering work with $LaTi_{0.15}Cu_{0.85}O_3$, the same catalyst used in Reference [41], was carried out to eliminate gallic acid, a primary intermediate of benzoic acid oxidation [43]. The role of different operating variables was studied. Whereas the catalyst and ozone doses exerted a positive influence in the ozonation rate, the increment in the initial acid gallic concentration diminished the conversion. The activity of the catalyst in terms of acid elimination was kept for consecutive cycles. However, the catalyst displayed a partial deactivation in terms of total organic carbon (TOC) elimination after the second reuse. In any case, the TOC degree was still higher than the one achieved in the non-catalytic system.

Finally, Carbajo et al. [44] went a step further in analysing the activity of the same catalyst, $LaTi_{0.15}Cu_{0.85}O_3$, in the ozonation of four real phenolic wastewaters coming from agro-industrial field, a wine distillery industry, olive debittering and from olive oil production. The main goal was to study the activity and stability of the catalyst together with the influence of the different operating variables. The results suggested that if enough time was allowed the catalytic ozonation of the phenolic mixture achieved 100% of mineralization. Moreover, the increment of temperature promoted the mineralization level.

Some of the compounds that cannot be easily removed from drinking water or wastewater by classical treatments are pharmaceutical compounds. Catalytic ozonation allows high removal of organic carbon of these compounds being the most appropriate process. As a first approach Beltran et al. [45] tested two copper and cobalt perovskites, $LaTi_{0.15}Cu_{0.85}O_3$ and $LaTi_{0.15}Co_{0.85}O_3$, as catalysts to remove in the presence of ozone sulfamethoxazole, a synthetic antibiotic usually found in municipal wastewaters. Some experiments were also carried out in the presence of activated carbon, as promoter of the activation of ozone. The results showed that catalytic or promoted ozonation were not necessary to eliminate sulfamethoxazole from water, because it can be removed only by the action of ozone. However, from a practical point of view, the combined ozone processes are clearly recommended in order to remove the resulting total organic carbon (TOC). The catalytic ozonation of two pharmaceutical compounds, the drug diclofenac and the synthetic hormone 17-ethynylstradiol, was also conducted on the same perovskites by the same authors [46], obtaining similar results to those observed in their previous work. Both compounds can be eliminated by direct ozonation; however, when copper perovskite was used, the TOC removal reached the 90% after 2 h of reaction.

To date, the main advantage of catalytic ozonation is the ability to improve the mineralization degree achieved at the end of the process. In this sense, non-substituted perovskites type $LaBO_3$ (B=Fe, Ni, Co and Mn) and substituted perovskites type $LaB_xCu_{1-x}O_3$ (B=Fe and Al), have been proposed as effective catalysts in ozonation processes of oxalic acid and dye C.I. Reactive Blue 5 [47]. Most of perovskites tested showed better performances in the catalytic ozonation of oxalic acid than single ozonation. On the contrary, in the case of the removal of dye, conversion values reached through single ozonation were slightly higher than in catalysed systems. However, the results in terms of TOC removal were better in the presence of catalyst and $LaCoO_3$ allowed almost complete mineralization of dye after 3 h of reaction under the conditions tested. A key factor in the removal of contaminants seems to be the presence of lattice oxygen vacancies, which are able to activate adsorbed species.

In general, the catalytic activity seems to be enhanced by high surface area and easy access of reactants to the active sites. Then high surface area could be of interest to improve the activity of catalysts. Concerning to that, Afzal et al. [48] studied the behaviour of high surface area perovskites for catalytic ozonation of 2-chlorophenol. A nanocasting technique (NC), using SBA-15 as a template, was employed for the synthesis of NC-LaMnO$_3$ and NC-LaFeO$_3$ catalysts, with high surface area. Authors compared these catalysts with the same perovskites synthesized by conventional citric acid (CA) assisted route, CA-LaMnO$_3$ and CA-LaFeO$_3$, as well as with Mn$_3$O$_4$ and Fe$_2$O$_3$. They found that NC-perovskites, containing easily accessible active sites, showed higher catalytic activity (80% of TOC removal) than their counterpart CA-perovskites (35% of TOC removal). Mn$_3$O$_4$ and Fe$_2$O$_3$ were the worst catalysts.

Bromide, Br−, is usually present as micro-emerging pollutant in some water matrixes to be degraded. The ozonation treatment leads to the formation of bromate, BrO$_3^-$, which is considered a potential carcinogen to humans by the World Health Organization, therefore being necessary its removal from the water. Zhang et al. [49] tested two perovskites, LaCoO$_3$ and LaFeO$_3$, as catalysts for the simultaneous removal of BrO$_3^-$ and benzotriazole (BZT) in the presence of ozone. 71% of BrO$_3^-$ was eliminated and BZT was completely degraded in only 15 min when LaCoO$_3$ was used. LaFeO$_3$ resulted in being inactive for BZT degradation; however, it reduced BrO$_3^-$ to HOBr/OBr$^-$ efficiently. The surface hydroxyl groups present in both perovskites were key in the involved reactions.

When the catalytic ozonation does not lead to a high degree of mineralization, it is necessary to combine it with other oxidation systems. The integration of catalytic ozonation and photocatalysis seems to be the most appropriate solution [50]. Photocatalytic ozonation is an advanced ozonation route allowing high removal of organic carbon by combining the beneficial effects of ozonation with the generation of hydroxyl radicals via electron hole formation (free radical oxidation). Considering that perovskites have been successfully used in catalytic ozonation, some authors studied the O$_3$/UVradiation/perovskites system. Thus, Rivas et al. [51] carried out the advanced oxidation of pyruvic acid in presence of LaTi$_{0.15}$Cu$_{0.85}$O$_3$ with several oxidation systems: O$_3$, UV radiation, O$_3$/UV radiation, O$_3$/perovskite, UV radiation/perovskite, O$_3$/UV radiation/perovskite, H$_2$O$_2$/UV radiation, H$_2$O$_2$/UV radiation/perovskite, being O$_3$/UV radiation/perovskite the system investigated in more detail. The efficiency of the oxidation systems was examined in terms of the economic cost as a function of removal percentage of pyruvic acid and TOC. As expected, ozone was not able of eliminating pyruvic acid, displaying conversion values around 20% after 3 h of reaction. In contrast, application of UV radiation led to 40% of pyruvic acid elimination. Finally, for the combined O$_3$/UV radiation/perovskite system, the pyruvic acid removal reached 100%, while the mineralization degree obtained was 80%. Therefore, under the operating conditions investigated, the photocatalytic ozonation seems to be the best option.

3. Processes Based on Hydrogen Peroxide

3.1. Fenton-like Reactions (H$_2$O$_2$/catalyst)

The homogeneous Fenton system implies the reaction of Fe^{2+} with H$_2$O$_2$ to generate hydroxyl radicals (Equation (4)), with a high reactivity and high oxidant power, capable of oxidize organics according to Equations (1)–(3).

$$Fe^{2+} + H_2O_2 \rightarrow Fe^{3+} + HO\bullet + HO^- \tag{4}$$

The generated HO• radicals can re-combine with Fe^{2+}:

$$Fe^{2+} + HO\bullet \rightarrow Fe^{3+} + HO^- \tag{5}$$

The ferric ions formed may decompose the hydrogen peroxide into water and oxygen, following the Equations (6)–(10), in which ferrous ions and radicals are also generated. The reaction of H$_2$O$_2$ with Fe^{3+} is referred in literature as Fenton-like reaction [52].

$$Fe^{3+} + H_2O_2 \rightarrow Fe\text{-}OOH^{2+} + H^+ \tag{6}$$

$$\text{Fe-OOH}^{2+} \rightarrow \text{HO}_2\bullet + \text{Fe}^{2+} \qquad (7)$$

$$\text{Fe}^{2+} + \text{HO}_2\bullet \rightarrow \text{Fe}^{3+} + \text{HO}_2^- \qquad (8)$$

$$\text{Fe}^{3+} + \text{HO}_2\bullet \rightarrow \text{Fe}^{2+} + \text{O}_2 + \text{H}^+ \qquad (9)$$

$$\text{HO}\bullet + \text{H}_2\text{O}_2 \rightarrow \text{H}_2\text{O} + \text{HO}_2\bullet \qquad (10)$$

Other reactions involving radicals in the Fenton process are:

$$\text{H}_2\text{O}_2 + \text{HO}\bullet \rightarrow \text{HO}_2\bullet + \text{H}_2\text{O} \qquad (11)$$

$$\text{HO}_2\bullet + \text{HO}_2\bullet \rightarrow \text{H}_2\text{O}_2 + \text{O}_2 \qquad (12)$$

Notice that by reaction (11) H_2O_2 acts as sink for HO•, diminishing the oxidizing power of the Fenton reactants.

The homogeneous Fenton system, which implies the reaction of Fe^{2+}/Fe^{3+} in solution with H_2O_2, has several drawbacks. By one hand, the chemical reactivity of iron is strictly dependent on the pH and only at pH $\approx$ 3, all three Fenton-active species of Fe^{2+}, Fe^{3+} and $Fe(OH)^{2+}$ coexist together. On the other hand, the final effluent contains high metal concentrations, which have to be recovered by additional treatment. In the heterogeneous systems Fe^{2+} and Fe^{3+} are part of a solid, which results in different advantages, especially related to recovery of the catalyst and the low leaching of ions.

The Fenton-like reactions involved in a heterogeneous system are the following:

$$\text{S-Fe}^{2+} + \text{H}_2\text{O}_2 \rightarrow \text{S-Fe}^{3+} + \text{HO}\bullet + \text{HO}^- \qquad (13)$$

$$\text{S-Fe}^{3+} + \text{H}_2\text{O}_2 \rightarrow \text{S-Fe}^{2+} + \text{HO}_2\bullet + \text{H}^+ \qquad (14)$$

$$\text{S-Fe}^{2+} + \text{HO}\bullet \rightarrow \text{S-Fe}^{3+} + \text{HO}^- \qquad (15)$$

$$\text{S-Fe}^{3+} + \text{HO}_2\bullet \rightarrow \text{S-Fe}^{2+} + \text{O}_2 + \text{H}^+ \qquad (16)$$

$$\text{Organic} + \text{HO}\bullet \rightarrow \text{R}\bullet + \text{H}_2\text{O} \rightarrow \ldots \rightarrow \text{CO}_2 + \text{H}_2\text{O} \qquad (17)$$

where S represents the surface of the catalyst. The reaction (14) is rate-limiting since its rate constant is ca. four orders of magnitude lower than that of reaction (13).

Although the classical Fenton system is based on the use of Fe^{2+}/H_2O_2, other elements with multiple redox sates (like chromium, cerium, copper, cobalt, manganese and ruthenium) can directly decompose H_2O_2 into HO• through conventional Fenton-like pathways [53]. Therefore, perovskites containing these elements, mainly in B position, can be used in this kind of reaction.

Rhodamine B (RhB) is one of the most studied organic pollutants in water in the Fenton-like reactions [54–58]. The first study reporting the use of perovskites as catalysts for the removal of RhB in Fenton-like reactions was carried out by Luo et al. [54]. In this work, $BiFeO_3$ magnetic nanoparticles (BFO MNPS) prepared by sol-gel method were tested in the degradation of RhB in the presence of H_2O_2 at 25 °C and pH = 5. According to the isoelectric point of BFO MNPs (I.P. = 6.7), under these conditions, the anionic form of the dye (pKa = 3.7) interacts easily via electrostatic forces with the positively charged catalyst particles. By selecting initial H_2O_2 and catalyst concentrations as 10 mM and 0.5 g/L, respectively, 95.2% of RhB was degraded in 90 min and a TOC removal of 90% was achieved within 2 h, in contrast to the removal of only 10% of RhB in the presence of Fe_3O_4 nanoparticles. By Montecarlo (MC) simulations authors concluded that after the adsorption of H_2O_2 molecules on the surface hollow sites of BFO MNPs facets, they are activated to generate HO• radicals, which then decompose RhB into other smaller organic compounds and CO_2. BFO MNPs showed excellent chemical stability during reaction (as checked by XPS), being reusable for at least five cycles, without a significant loss of activity. BFO MNPs were also tested in the degradation of methylene blue and phenol, leading to 79.5% and 82.1% of removal, respectively.

Zhang et al. [55] synthesized a series of Cu-doped LaTiO$_3$ perovskite (LaTi$_{1-x}$Cu$_x$O$_3$, x = 0.0–1.0) by a sol-gel method, which resulted be very efficient for the degradation of RhB with H$_2$O$_2$ in a pH range of 4–9. In contrast to the absence of activity of sample containing only titanium, the coexistence of Ti^{3+}/Ti^{4+} and Cu$^+$/Cu^{2+} in the perovskite structure of partially substituted samples allowed the degradation of 8 mg/L of RhB through redox cycles involving the transformation of H$_2$O$_2$ into HO$\bullet$ and HO$_2\bullet$/O$_2\bullet^-$. For a H$_2$O$_2$ concentration of 10 mM, about 84% of RhB was decolorized within 2 h in the presence of 1.4 g/L of LaTi$_{0.4}$Cu$_{0.6}$O$_3$. Notice that the amount of RhB degraded was slightly lower than the observed by Luo et al. [54]; although the catalyst amount used in Reference [55] was almost three times higher, the initial concentration of RhB was approximately the double. The reduction of H$_2$O$_2$ to O$_2$, which is carried out by oxygen vacancy [22], was not observed in this reaction.

The surface area of perovskites is low and as a consequence, the interaction between the contaminants and the active sites is limited. In order to improve the catalytic efficiency of perovskite-like oxides by increasing their surface area, some strategies have been developed, such as their supporting on mesoporous silica supports [56–58] or in honeycombs [59] and the formation of nanocomposites [60,61].

In this sense, La-FeO$_3$/SBA-15 [56] was more efficient than non-supported LaFeO$_3$ for catalysing RhB oxidation in the presence of H$_2$O$_2$ under ambient conditions due to a synergic effect between the large capacity of mesoporous SBA-15 for RhB adsorption and the high number of active sites exposed in LaFeO$_3$ nanoparticles for reacting with H$_2$O$_2$. The best catalyst was the sample containing many oxygen vacancies (as deduced from XPS results), which are a key factor influencing the performance of these catalysts in oxidation reactions. The catalyst was efficient in a wide pH range (2–10). Under the optimum conditions, a degradation of RhB of 87% was achieved after 3 h. No leaching of Fe^{3+} was observed in the solution after reaction, the contribution of homogeneous Fenton reaction being discarded. The stability of La-FeO$_3$/SBA-15 was also confirmed by carrying out four cycles of reutilization, which showed no deactivation of the sample. The catalyst was also applied for the degradation of other organic dyes, achieving a decomposition of 66% for methylene blue and 42% for brilliant red X-3B and direct scarlet 4BS.

The good synergy between the support and the LaFeO$_3$ perovskite was explored by the same authors [58], who tested different supports based on mesoporous silica, such as SBA-15, SBA-16 and MCF and on nanosized silica powders (NSP). Different factors influence on the catalytic behaviour for degradation of RhB. By one hand, the RhB adsorption on the support is a crucial step of the reaction and as a result, the combination of LaFeO$_3$ with a non-porous support showing a low capacity of adsorption decomposed the RhB in a little extent. On the other hand, a network of pores with short length is necessary to allow the transportation of RhB to the active sites of LaFeO$_3$. In this sense, the shorter the pore length, the faster the RhB molecules reached the catalytic centres and were oxidized (see the transport process in Figure 2). Authors concluded that LaFeO$_3$ supported on MCF containing randomly distributed pores with short length was the best catalyst for oxidative degradation of RhB in aqueous solution, achieving a removal of the contaminant of 97% in 2 h.

SBA-15 was also used by the same authors as support of a perovskite-type oxide La$_2$CuO$_4$ containing a few amounts of CuO [57]. The solid was tested in the degradation of RhB and organic dyes, including reactive brilliant red X-3B, direct scarlet 4BS and methylene blue under ambient conditions. The catalyst was active in a wide pH range (2–10) and depletion of RhB between 85% and 95% was produced after 3 h, depending on the amount of catalyst. The mineralization of RhB into CO$_2$ was completed and the catalyst could be recycled. Although the activity decreased in ca. 14% in the fifth cycle, it could be recovered after a treatment of the used catalyst in air at 500 °C for 2 h.

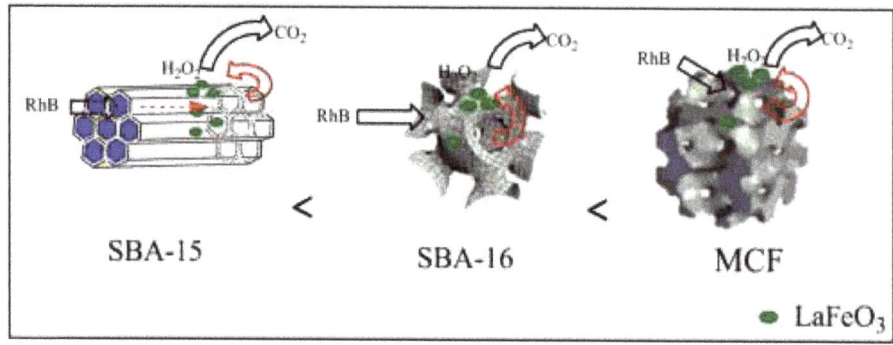

Figure 2. A proposed scheme of transporting RhB from the solution to the pore and then to the surface-active site over LaFeO$_3$ catalysts supported on porous SBA-15, SBA-16 and MCF. With permission from [58].

Another approach for modifying the surface properties and reactivity of perovskites is the formation of nanocomposites [60,61]. In this regard, a novel 3D perovskite-based composite BiFeO$_3$/carbon aerogel (BFO/CA) prepared by sol-gel method led to a 95% of degradation of ketoprofen in 150 min and a TOC removal of 60% after 5 h [60]. These activities values were significantly higher than those obtained for bulk BFO and nano BFO, due to the higher reducibility of Fe^{3+} and Co^{3+} species in the composite, as deduced from TPR studies and to the dispersion of active sites not only on the surface of CA support but along the 3D structure of CA. Furthermore, the catalyst was active in a wide pH range of 3–7 and the leaching of iron was low.

A La$_{1+x}$FeO$_3$ (L$_{1+x}$FO, $0 \leq x \leq 0.2$) nanocomposite formed between LaFeO$_3$ and an inert La$_2$O$_3$, resulted to be twice more active for degradation of methyl orange that the pristine LaFeO$_3$ [61]. The modification of surface properties, such as surface Fe^{2+} concentration, surface defects, H$_2$O$_2$ adsorption capacity and charge-transfer rate led to an enhanced Fenton-like activity in the composite. The most notorious aspect of this work was that the major reactive species were not hydroxyl radicals but singled oxygen (^{1}O$_2$), as deduced from in situ electron paramagnetic resonance analysis and radical scavenging experiments. Authors proposed the corresponding mechanism of ^{1}O$_2$-based composite/H$_2$O$_2$ system. 100% of contaminant was degraded in 90 min at pH = 3 and a total organic carbon (TOC) removal of 96% was achieved after 4 h.

Other contaminants degraded by LaFeO$_3$ perovskite, in this case auto supported, were different pharmaceutical and herbicides [62]. Among them, sulfamethoxazole (SMX) was completely removed in LaFeO$_3$–H$_2$O$_2$ system after 2 h at neutral pH. By formation of a surface complex between LaFeO$_3$ and H$_2$O$_2$, the O–O bond in H$_2$O$_2$ is weakened and chemical environment of iron changes, the Fe^{3+}/Fe^{2+} redox potential decreasing significantly, which accelerates the cycle of Fe^{3+}/Fe^{2+} and produces more HO• and O$_2$•$^-$/HO$_2$• radicals, enhancing the Fenton-like removal of organic compounds. The TOC removal was 22% in 2 h and SMX was transformed into simpler aliphatic acids, mostly biodegradable.

Due to its abundance in most of wastewater effluents and its toxicity, phenol is a usual organic compound model in developing methods for water remediation, including AOPs. The removal of phenol and phenolic compounds has been tested in Fenton-like reactions on different perovskites, mainly containing iron or copper in B position [63–65]. LaFeO$_3$ and BiFeO$_3$ were tested by Rusevova et al. [63] in the degradation of phenol. The influence of reaction temperature, catalyst and H$_2$O$_2$ concentrations and pH, on the catalytic behaviour was studied. The rate constant for phenol degradation, which increased with temperature, was 3-fold higher when initial reaction pH diminished from 7 to 5. Conversion values of phenol of 90–95% were achieved after 6 h and leaching of metals was negligible. The most new-fangled aspect of the study was that in order to settle the nature of the active oxidizing species authors used compound specific stable isotopic analysis (CSIA) as alternative to other conventional techniques. Based on their results, authors concluded that the major species involved in phenol degradation were hydroxyl radicals. They extended the application of both LaFeO$_3$ and BiFeO$_3$ to the removal of methyl terc-butyl ether (MTBE), for which a depletion of 80% was obtained after 6 h.

Different perovskite-like oxides LaBO$_3$ (B = Cu, Fe, Mn, Co, Ni) synthesized using the Peccini method were tested in Fenton-like degradation of phenol but only LaCuO$_3$ and LaFeO$_3$ were active [64]. Authors studied the recyclability of the catalysts during 3 cycles for LaCuO$_3$ and 40 cycles for LaFeO$_3$. The induction period observed in the first cycle for LaFeO$_3$ was significantly shortened for the second and successive cycles. In this sense, a degradation of phenol of 75% was produced in 5 h in the second cycle, in contrast to the 22% observed in the first one. The reasons for this improvement in the activity were an increase in the surface concentration of oxygen containing species (water and carbonate) involved in the transformation and the formation of dispersed particles of iron oxides on the surface. The TOC conversion of 21–22% after 10 h did not change for the different cycles.

Hammouda et al. [65] prepared ceria perovskite composites CeO$_2$-LaCuO$_3$ and CeO$_2$-LaFeO$_3$, which were more active for the degradation of bisphenol than non-doped perovskites, especially at short reaction times. Furthermore, CeO$_2$ improved the stability of perovskites towards leaching of metals. Authors attributed the enhancement in the activity to the fact that, as observed by XPS, more Ce^{3+} ions were formed in the ceria-perovskite catalysts, due to an electron transfer from the transition metal of perovskites to the CeO$_2$. As a result, more oxygen radicals were formed by interaction of H$_2$O$_2$ with Ce^{3+}, favouring the Fenton-like degradation of the contaminant. By following the evolution of the intermediates formed, authors proposed a mechanism of reaction and a degradation pathway.

In order to improve the catalytic ability of BiFeO$_3$ nanoparticles to degrade recalcitrant pollutants, some authors have proposed an in-situ surface modification by using chelating agents [66,67]. In this regard, the bisphenol A (BPA) degradation in a wide pH range (4–9) was accelerated when the nano-BiFeO$_3$ were modified by adding different ligands to the Fenton solution, such as tartaric acid, formic acid, glycine, nitrilotriacetic acid and ethylenediaminetetraacetic acid (EDTA) [66]. EDTA was the most efficient chelating agent, mainly because of a higher HO• formation from the H$_2$O$_2$ decomposition. Under the optimum conditions 91.2% of BPA was removed within 2 h, in contrast to the 20% of BPA degraded with unmodified BFO. Although the use of chelating agents increased the contribution of Fenton homogeneous reaction by formation of soluble iron complexes, the trend observed in the BPA degradation for reactions carried out with different ligands did not follow the order of leached ions, indicating the irrelevant contribution of homogenous reaction to BPA degradation. As EDTA was the most efficient chelating agent, it was also used by same authors [67] as ligand for BiFeO$_3$ in the degradation of triclosan (5-chloro-2-(2,4-dichlorophenoxy)phenol), a broad-spectrum antibacterial agent widely used in personal and health care products. When pristine BFO were used, triclosan was mainly transformed into 2,4-dichlorophenol, a carcinogenic compound. The addition of EDTA modified significantly the dechlorination ratio of triclosan, which increased from 26.4% for H$_2$O$_2$-BFO sample up to 97.5% in the chelated system. Triclosan was degraded almost completely in 3 h under the optimal conditions.

Another strategy to improve catalytic activity of perovskites in AOPs is the hetero-doping in order to produce more active sites of the low-valence B-site transition metals (i.e., Fe^{2+}, Cu$^+$ and Ti^{3+}) or to introduce oxygen vacancies, which can facilitate the transformation of H$_2$O$_2$ into HO• [39,40]. In this sense, some perovskites containing partially substituted manganese in B position, have been tested in Fenton-like reactions for the degradation of methylene blue (MB) [68], different dyes [69] and paracetamol [70].

Maghalaes et al. [68] tested LaMn$_{1-x}$Fe$_x$O$_3$ and LaMn$_{0.1-x}$Fe$_{0.9}$Mo$_x$O$_3$ perovskites in the decomposition of H$_2$O$_2$ to O$_2$ and in the oxidation of MB. The presence of manganese in the perovskites seemed to play an important role on the H$_2$O$_2$ decomposition rate, which decreased with the amount of Mn substituted by Fe and/or Mo. However, LaMnO$_3$ was not active for the MB discoloration, which suggested that it was able to transform the H$_2$O$_2$ into O$_2$ but it was unable to form the HO• radicals, necessary to degrade the dye molecules. On the contrary, samples substituted by Mo degraded MB up to 20% in 1 h.

Jahuar et al. [69] synthesized a series of manganese-substituted lanthanum ferrites having compositions LaMn$_x$Fe$_{1-x}$O$_3$ (x = 0.1–0.5) by a sol–gel auto-combustion method, which were used as catalysts in the removal of anionic dyes (Remazol Turquoise Blue, Remazol Brilliant Yellow) and cationic dyes (MB, Safranine-O) by the action of H_2O_2, in the absence and presence of visible-light. The initial pH of solution was fixed in all cases to the value of 2. Unsubstituted LaFeO$_3$ produced a low dye degradation for long time periods, exhibiting a poor catalytic activity under dark conditions. However, the partial substitution of iron by manganese led to catalysts able to degrade over 90% of dye in time periods of 150–300 min, due to the Fenton-like activity of manganese ions, capable of existing in various oxidation states. In the presence of light, an enhancement in the catalytic activity was produced and degradation times were reduced to 25–70 min.

The contribution of manganese ions to Fenton-like reaction was, on the contrary, discarded by Carrasco-Díaz et al. [70] in the decomposition of paracetamol by H_2O_2 under mild reaction conditions (25 °C and pH ≈ 6) in the presence of LaCu$_x$M$_{1-x}$O$_3$ (0.0 ≤ x ≤ 0.8, M = Mn, Ti) perovskite-like oxides prepared by amorphous citrate decomposition. Degradation values of paracetamol between 80% and 97% were achieved after 5 h. XPS studies of the catalysts allowed authors to conclude that Cu^{2+}/Cu^+ were the catalytically active species, the catalysts containing a higher amount of copper at the surface, mainly as Cu^{2+}, being the most active. The titanium and manganese species seemed not to be responsible of the enhanced activity observed in some of the substituted samples with respect to that of LaCuO$_3$. The catalysts were recyclable for at least three cycles and a negligible leaching of metals was produced. TOC values of 47–54% were achieved.

Finally, some mathematical analysis of the heterogeneous oxidations of contaminants by perovskites have been carried out. More concretely, mathematical modelling of photo-Fenton-like oxidation of acetic acid by LaFeO$_3$ has been reported [71,72]. From the experimental results authors concluded that the main reactions occurring in the system were the complete mineralization of acetic acid by H_2O_2 due to the presence of the catalyst and the decomposition of H_2O_2 into water and O_2 in the homogeneous phase. Therefore, this kind of reaction should not be considered as an AOP, because no hydroxyl radicals were formed.

Table 1 summarizes the conversion values and reaction conditions for the use of perovskites in the degradation of different organics by Fenton-like reactions and photo Fenton-like reactions, these last being revised in the following section.

Table 1. Oxides type perovskite used in Fenton-like and photo Fenton-like reactions of organic pollutants in aqueous solution.

Reference	Catalyst	Target Pollutant	Concentration Catalyst	Concentration Pollutant	H_2O_2	Treatment Efficiency Degradation of Pollutant	TOC Removal
[54]	$BiFeO_3$	RhB, phenol, MB	0.5 g/L	4.79 g/L	10 mM	95.2% (RhB) in 90 min	90% in 2 h
[55]	$LaTi_{1-x}Cu_xO_3$ (x = 0.0–1)	RhB	1.4 g/L	8 g/L	10 mM	84% in 2 h	-
[56]	$LaFeO_3$/SBA-15	RhB, MB, brilliant red X-3B, direct scarlet 4BS	2 g/L	9.1 mg/L	0.34 mM	87% in 3 h (RhB) 66% MB 42% for the others	-
[57]	La-Cu-O/SBA-15	RhB and other dyes	Not indicated, [RhB]/[catalyst]= 0.0045, 0.077	-	-	85–95% in 3 h	100% in 3 h
[58]	$LaFeO_3$/SBA-15, SBA-16, MCF and non-porous silica	RhB	2 g/L	9.6 mg/L	0.34 mM	97% in 2 h for $LaFeO_3$/MCF	-
[60]	$BiFeO_3$/carbon aerogel (BFO/CA)	ketoprofen	0.3 g/L	40 mg/L	12 mM	95% in 150 min	60% in 5 h
[61]	$La_{1+x}FeO_3$ ($L_{1+x}FO$, $0 \leq x \leq 0.2$) nanocomposite	Methyl orange	0.5 g/L	5 mg/L	0.198 M	100% in 90 min (pH 3)	96% in 4 h
[62]	$LaFeO_3$	Different herbicides and pharmaceutical	1.4 g/L	3 mg/L	23 mM	100% of SMX in 2 h (pH 6.5)	22% in 2 h
[63]	$LaFeO_3$, $BiFeO_3$	phenol, MTBE	0.01–1 g/L	25 mg/L phenol 50 mg/L MTBE	3 g/L = 88 mM	90–95% phenol in 6 h 80% MTBE in 6 h	-
[64]	$LaBO_3$ (B= Cu, Fe, Mn, Co, Ni)	phenol	5 g/L	0.01 M	0.7 M	85% in 10 h	21% in 10 h
[65]	Ceria-$LaCuO_3$, ceria-$LaFeO_3$	Bis-phenol	0.2 g/L	20 mg/L	10 mM	98% in 45 min (ceria-$LaCuO_3$) 92% in 42 min (ceria-$LaFeO_3$)	-
[66]	$BiFeO_3$ modified by chelating agents	BPA	0.5 g/L	0.1 mM	10 mM	91.2% in 2 h	-
[67]	$BiFeO_3$ (BFO); BFO modified by EDTA	Triclosan	0.5 g/L	34.5 mM	10 mM	82.7% in 3 h (BFO) 100% in 30 min (BFO-EDTA)	-
[68]	$LaMn_{1-x}Fe_xO_3$, $LaMn_{0.1-x}Fe_{0.9}Mo_xO_3$	MB	30 mg (volume not indicated)	0.1 g/L	2.9 mM	20% in 1 h ($LaMn_{0.01}Fe_{0.9}Mo_{0.09}O_3$)	-
[69]	$LaMn_xFe_{1-x}O_3$ (x = 0.1–0.5), absence and presence of visible light	Anionic and cationic dyes	0.2 g/L	15 mg/L cationic dyes 60 mg/L anionic dyes	17 mM	Anionic dyes: 66–98% in 4 h (no light); 90–95.8% in 70 min (light) Cationic dyes: 90–99.4% in 5 h (no light); 98–99.7% in 30 min (light)	-

Table 1. Cont.

Reference	Catalyst	Target Pollutant	Concentration Catalyst	Concentration Pollutant	H_2O_2	Treatment Efficiency Degradation of Pollutant	Treatment Efficiency TOC Removal
[70]	$LaCu_xM_{1-x}O_3$ ($0.0 \leq x \leq 0.8$, M= Mn, Ti)	paracetamol	0.2 g/L	50 mg/L	13.8 mM	80–97% in 5 h	47–54% in 5 h
[73]	$LaMnO_3$ + UV light	phenol	0.6 g/L	0.1 g/L	14.8 mM	99.92% in 4 h	-
[74]	Catalyst ($Bi_{0.97}Ba_{0.03}FeO_3$, $BiFe_{0.9}Cu_{0.1}O_3$, $Bi_{0.97}Ba_{0.03}Fe_{0.9}Cu_{0.1}O_3$) + visible light	2-chlorophenol	0.4 g/L	50 mg/L	10 mM	100% in 70 min ($BiFe_{0.9}Cu_{0.1}O_3$, $Bi_{0.97}Ba_{0.03}Fe_{0.9}Cu_{0.1}O_3$) 63% in 70 min ($Bi_{0.97}Ba_{0.03}FeO_3$)	68% in 1 h ($BiFe_{0.9}Cu_{0.1}O_3$) 73% in 1 h ($Bi_{0.97}Ba_{0.03}Fe_{0.9}Cu_{0.1}O_3$)
[75]	$EuFeO_3$ (EFO) + visible light	RhB	1 g/L	5 mg/L	0.2 mM	37% EFO + vis. light 50% EFO + H_2O_2 71% EFO + H_2O_2 + vis. light	-
[76]	$LaFe_{1-x}Cu_xO_3$ + visible light	Methyl orange	0.8 g/L	10 mg/L	8.8 mM	92.9% in 1 h	-
[59]	$LaFeO_3$ + UV light $Pt/LaFeO_3$ + UV light	tartrazine	81.25 g/L	40, 60, 80 mg/L	0.0019–0.0076 mol/h	43% in 3 h ($LaFeO_3$) 63% in 3 h ($Pt/LaFeO_3$) (40 mg/L and 0.0038 mol/h) 100% in 30 min (catalyst + UV light)	45% in 3 h ($LaFeO_3$); 65% in 3 h ($Pt/LaFeO_3$) (40 mg/L and 0.0038 mol/h) 100% in 40 min (catalyst + UV light)
[77]	$LaMeO_3$ (Me= Mn, Co, Fe, Ni, Cu)/cordierite, and $Pt/LaMnO_3$/cordierite	acetic acid	9.4–40 g/L	1.26 g/L	83 mM	-	54% in 5 h ($Pt/LaMnO_3$) 60% in 5 h ($LaFeO_3$)
[78]	$LaFeO_3$/corundum with different loads of $LaFeO_3$	acetic acid, ethanol, acetaldehyde, oxalic acid	93.75 g/L	0.5 g/L	83 mM	-	97% (acetic acid); 53% (ethanol); 62% (acetaldehyde) and 95% (oxalic acid) in 5 h
[79]	$LaFeO_3$/corundum	MTBE	9.4–4 g/L	0.5 g/L	0–42 mM	-	100% in 2 h
[80]	$LaFeO_3$, $BiFeO_3$, $LaTi_{0.15}Fe_{0.85}O_3$ and $BiTi_{0.15}Fe_{0.85}O_3$	methylparaben	0.1 g/L	5 mg/L	0.5 mM	82.8% in 90 min	-
[81]	graphene-$BiFeO_3$ + EDTA + visible light	TTBBA	0.5 g/L	0.011 g/L	20 mM	80% in 15 min	62.8% in 3 h

RhB: Rhodamine B; MB: Methylene Blue; MTBE: methyl tert-butyl ether; BPA: bis-phenol A; TTBBA: tetrabromobisphenol A.

3.2. Photo Fenton-Like Reactions (H_2O_2/Catalyst/Light)

In the photocatalytic oxidation processes, the electron–hole pairs in the catalyst are produced via the irradiation of the UV light and the oxidative radicals are formed between the catalyst and water interface [82]. The formation rate of HO• radicals in photo-Fenton processes is higher than in Fenton processes. While the Fenton reaction is governed principally by Equation (4) leading to the formation of HO• radicals, in the Photo-Fenton process occurs, in addition, the photolysis of H_2O_2:

$$H_2O_2 + h\nu \rightarrow 2\ HO\bullet \tag{18}$$

and the photo reduction of Fe^{3+}:

$$Fe^{3+} + H_2O + h\nu \rightarrow HO\bullet + Fe^{2+} + H^+ \tag{19}$$

Different perovskite oxides, non-supported [69,73–76], supported on monoliths [59,77–80] or in form of composites [81], have been used as catalysts for the degradation of several organics in the presence of H_2O_2 under light irradiation conditions.

A $LaMnO_3$ perovskite prepared by co-precipitation method resulted to be an excellent photo-Fenton catalyst of oxidation of phenol [73]. The phenol conversion (99.92% in 4 h) obtained when $LaMnO_3$ was activated by UV radiation in the presence of stoichiometric amount of H_2O_2 necessary to degrade phenol, was even higher than the achieved when using TiO_2 as catalyst (98% in 4 h). Furthermore, $LaMnO_3$ could be regenerated by calcination after reaction, yielding to similar catalytic performance to that of the first cycle.

In Reference [75] a series of iron perovskites containing europium in B position, $EuFeO_3$ (EFO), calcined at different temperatures, was used for the photodegradation of RhB by combination of visible light and H_2O_2. By the action of visible light, electron-hole pairs were formed in EFO nanoparticles and electrons were easily trapped by H_2O_2, leading to the formation of HO• radicals. In addition, a complex between Fe^{3+} at the surface ($\equiv Fe^{3+}$) and H_2O_2 was formed and $\equiv Fe^{3+}$ was transformed into $\equiv Fe^{2+}$, generating HO• and $HO_2\bullet$, which decomposed RhB. Authors studied the effect of the calcination temperature of the catalysts on the band gap, microstructure and photocatalytic activity. The perovskite calcined at 750 °C showed the best catalytic behaviour, due to the combination of good crystallinity and appropriate BET surface area and band gap. The photodegradation of 37% of RhB after 3 h increased up to the 71% when H_2O_2 was added, which proves the Fenton-like activity of EFO nanoparticles.

One strategy to improve photocatalytic activity of ABO_3 perovskites is the substitution of the element in A or B position, which leads to the introduction of defects into the narrow band gap and to the formation of oxygen vacancies, which inhibit the recombination between the photogenerated electrons and the holes. In this regard, 2-chlorophenol (2-CP), was degraded in the presence of three metal doped $BiFeO_3$ (BFO) nanoparticles, H_2O_2 and visible light [74]. $BiFeO_3$ perovskite was substituted either in A position ($Bi_{0.97}Ba_{0.03}FeO_3$) or B position ($BiFe_{0.9}Cu_{0.1}O_3$) and in both ($Bi_{0.97}Ba_{0.03}Fe_{0.9}Cu_{0.1}O_3$). After only 70 min of visible light irradiation, Cu-doped BFO and Ba-Cu co-doped BFO almost completely removed 2-CP. The mineralization degree reached was of 68% and 73%, respectively. Authors concluded that in addition to the participation of Fe^{2+}/Cu^+ couple active for the formation of HO•, the oxygen vacancies on the surface can also participate by activating H_2O_2 molecules to form a lattice oxygen, which is furtherly desorbed as O_2.

More recently Phan et al. [76] studied the efficiency of $LaFeO_3$ (LFO) perovskites, doped with Cu in B position ($LaFe_{1-x}Cu_xO_3$), in the photo-Fenton decolorization of methyl orange (MO). Interestingly, the substitution of Cu into Fe-site in LFO modified the light absorption property of perovskite, as noted in the UV–vis absorption spectra and the corresponding band gap energy of LFO and $LaFe_{1-x}Cu_xO_3$ shown in Figure 3a,b, respectively. Notice that all the samples had suitable band gap energy for organic pollutant degradation under visible light irradiation. When Fe was substituted by 15 mol% of Cu

(LFO-15Cu), the MO degradation rate was improved in ca. 60% and 92.9% of MO was removed in only 1 h at an initial solution pH of 6. Under the optimum conditions, the photocatalytic performance of LFO-15Cu was also evaluated for two cationic dyes, rhodamine B (RhB) and methylene Blue (MB), obtaining even better results: 99.4% of degradation for RhB and 98.8% for MB in 60 min.

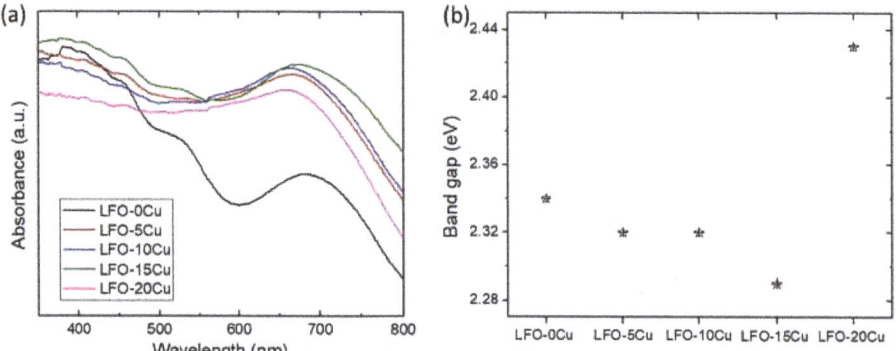

Figure 3. (a) UV–vis absorption spectra and (b) corresponding band gaps of LFO and $LaFe_{1-x}Cu_xO_3$. With permission from [76].

LaFeO$_3$ or Pt/LaFeO$_3$ perovskites supported on honeycomb monoliths have been tested in the degradation of tartrazine, a not-biodegradable dye used in food industries [59] by continuous flow of H_2O_2 in the presence of UV light at different pH values. The natural pH of solution (near 6) was the best operating condition, under which 100% of tartrazine was discoloured after 30 min of irradiation and mineralization was complete after 40 min. On the contrary, the discoloration was 50% under acidic condition (pH 3) and 55% under basic condition (pH 9), after 30 min of irradiation. In the first case, the excess of H$^+$ could react with HO• and produce water subtracting hydroxyl radicals necessary for the decomposition of tartrazine [83]. In the second one, H_2O_2 was accumulated in liquid medium because under alkaline conditions H_2O_2 has a very high stability [84] and as a consequence, the production of hydroxyls radicals was limited.

Supported LaMeO$_3$ (Me= Mn, Co, Fe, Ni, Cu) perovskites were prepared by impregnation of thin wall of monolithic honeycomb cordierite support with different active phase loadings and tested in the photo-Fenton oxidation of acetic acid [77]. In the case of LaMnO$_3$ sample, the honeycomb was also impregnated with 0.1% of Pt. Photo-Fenton activity was closely related to the amount of active phase supported on monolithic carrier and LaFeO$_3$ and Pt/LaMnO$_3$ perovskites were the best catalysts in terms of reaction rate. The addition of Pt enhanced the initial rate of acetic acid degradation, achieving the highest TOC removal, 18% in 1 h; however it did not enhance the catalytic performance after 5 h.

Different loads of LaFeO$_3$ perovskites supported over corundum monoliths were studied by the same authors in the photo-Fenton degradation of several organics [78] under UV irradiation. 97% of TOC removal was attained in the degradation of acetic acid after 4 h with the catalyst containing 10.64 wt% of LaFeO$_3$ and values of 53, 62 and 95% were achieved when ethanol, acetaldehyde and oxalic acid, respectively, were used as model pollutant.

The excellent catalytic behaviour of this catalyst was extended to other contaminants, as methyl terc-butyl ether (MTBE) [79]. About 100% of TOC removal and complete mineralization of MTBE into CO_2 and water was achieved if H_2O_2 was continuously dosed during irradiation time of 2 h at solution pH of 6.7. Although the TOC removal obtained by the combination of H_2O_2 and UV light in the absence of catalyst was quite high (about 97%), a very significant formation of CO was observed, indicating the importance of LaFeO$_3$ for the improvement in the mineralization of MTBE.

The degradation of methylparaben was studied in the presence of four ABO$_3$ perovskite catalysts (A: La, Bi and B: Fe, Ti-Fe) supported on a monolithic structure [80]. BiFeO$_3$ was the best catalyst and under the optimum conditions 82.8% of pollutant was degraded in only 90 min. In the absence of UV light, only 10% of methylparaben was removed. An interesting aspect of this work was the study of toxicity carried out by cress seed, showing an inhibition of only 1.09% in the growing of roots, which demonstrated the low toxicity of the products of degradation.

The combination of photocatalytic activity and oxidizing power of H_2O_2 was also applied for the degradation of tetrabromobisphenol A (TBBPA) with a graphene-$BiFeO_3$ composite as catalyst [81]. The catalytic activity was influenced by calcination temperature, pH, presence or not of EDTA, dosage of H_2O_2 and load of catalyst. The degradation of TBBPA approximately followed a kinetics of pseudo first order. Under the most favourable conditions and in the presence of EDTA the rate constant of TBBPA degradation with the graphene-$BiFeO_3$ was 5.43 times higher than that of $BiFeO_3$ and 80% of TBBPA was removed in 15 min. This enhancement in the catalytic activity was attributed to the increasing of the adsorption capacity (due to a large surface area) and to the high electron transfer ability of graphene in the composite, which favoured the generation of reactive species. The composite was stable and could be reused for five cycles without loss of catalytic activity.

3.3. Catalytic Wet Peroxide Oxidation (H_2O_2/Catalyst/Air)

Catalytic wet peroxide oxidation (CWPO) processes are based on the degradation of contaminants by the combined action of a solid catalyst, hydrogen peroxide and air in aqueous solution. In a certain extent, they are similar to Fenton-like processes, because they use H_2O_2 as oxidant; however CWPO processes are carried out in the presence of a flow of air or under pressurized air. They can work with high oxidation efficiency in a wide range of pH without leaching or production of sludge.

Apart from Fenton-type catalysts, including zeolitic materials or composite metal oxides, a reduced number of perovskites has also been tested for CWPO applications. It is important to note that very little theoretical and experimental information is available and there are only a few examples of CWPO using perovskites as heterogeneous catalysts.

The first application of a perovskite in CWPO was carried out by Ovejero et al. [85], who compared the activity of $LaTi_{0.45}Cu_{0.55}O_3$ with that of other catalysts containing Fe or Cu in the degradation of phenol. The reactions were carried out at 100 °C in a system pressurized with air at 1 MPa. The perovskite led to a complete elimination of phenol and a TOC removal of 92% in 45 min. The leaching of copper was 22%; however it was significantly lower than the leaching measured for other catalysts (between 64 and 74%). Considering the fact of the high stability of copper ions in perovskite structures, it is probable that most of the leached copper proceeded from La_2CuO_4 oxide, phase detected by XRD together with the perovskite phase. $LaTi_{0.45}Cu_{0.55}O_3$ was reused in a second cycle, the TOC removal decreasing only to 90%.

Three years later, the same authors extended the study of the CWPO degradation of phenol to other perovskites, in order to elucidate the influence of different reaction conditions (temperature, H_2O_2 peroxide concentration, catalyst concentration and air pressure) on the performance [86]. Three perovskites of $LaTi_{1-x}Cu_xO_3$ composition, with different substitution degree, were tested in the reaction. TOC removal values comprised between 88 and 94% were achieved at 100 °C under air pressure (1 MPa) and stoichiometric amount of H_2O_2 after 2 h. The temperature exerted a significant effect on the activity and only a 15% of TOC removal was reached when reaction was developed at 40 °C. The catalysts could be easily regenerated by calcination in air, leading to similar activity in the second run.

Less drastic operation conditions that the reported above were applied by Faye et al. [87] for the degradation of the same contaminant, phenol, by the action of several $LaFeO_3$ perovskites, synthesized by self-combustion method by varying the glycine/NO_3^- molar ratio. Thus, authors used a flow of air at atmospheric pressure and mild temperatures, 25 or 40 °C. Depending on synthesis conditions, strong differences in the structural, textural and reducibility characteristics were observed. The perovskite having the highest surface area exhibited the highest TOC abatement (76%) and very low iron leaching

(0.27 wt%) after 4 h of reaction at 40 °C. The perovskites were better catalysts than Fe_2O_3, for which a TOC removal of only 10% was reached after 4 h.

4. Processes Based on Peroxymonosulfate

4.1. Peroxymonosulfate Activated by Perovskites (PMS/catalyst)

The peroxymonosulfate ion (PMS) can be considered as a derivate of hydrogen peroxide by replacing one H-atom by a SO_3^- and is stable as triple potassium salt ($2KHSO_5 \cdot KHSO_4 \cdot K_2SO_4$), known by the commercial names of Oxone (from DuPont) and Caroat (from Evonik). PMS is quite stable in water solution and over a wide pH range and shows great potential for generating both sulphate and hydroxyl radicals. Sulphate radicals have a higher half-life than hydroxyl radicals (30–40 ls vs. 20 ns) and they are more selective to react with organics containing unsaturated bonds or aromatic rings during electron transfer, having some operation advantages. Recently, a review of AOPs based on sulphate radicals generated from PMS and persulfate has been published [88]. PMS in an unsymmetrical oxidant that, in the absence of any activator, can partially oxidize some organic compounds according to the redox potential of 1.82 V of reaction (Equation (20)).

$$HSO_5^- + 2H^+ + 2\,e^- \rightarrow HSO_4^- + H_2O \tag{20}$$

In order to generate sulphate radical, the decomposition of PMS must be carried out in presence of an activator, such as transition metals, ultraviolet irradiation (UV), microwave (MW), ultrasound (US), electron conduction and homogeneous and heterogeneous catalysts. The different methods for activation of PMS as well as their application for the removal or persistent organics have been recently revised by Ghanbari et al. [89]. Several nanostructured oxides and carbon materials have been tested as heterogeneous catalysts for PMS activation [90,91]. In spite of the high reactivity of nanostructured oxides, their low recyclability is a problem to be solved. Carbon materials show in general low activities and stabilities. Among heterogeneous catalysts used for PMS activation, materials based on cobalt play an important role. Thus, cobalt oxides, Co-metal oxide and Co-carbon-based supports have been applied for the degradation of different pollutants in presence of PMS. The mechanism for oxidation of organics by Co-assisted decomposition of PMS has been proposed as follows [92–94]:

$$Co^{2+} + HSO_5^- \rightarrow Co^{3+} + SO_4\bullet^- + OH^- \tag{21}$$

$$Co^{2+} + H_2O \rightarrow CoOH^+ + H^+ \tag{22}$$

$$CoOH^+ + HSO_5^- \rightarrow SO_4\bullet^- + CoO^+ + H_2O \tag{23}$$

$$CoO^+ + 2H^+ \rightarrow Co^{3+} + H_2O \tag{24}$$

$$Co^{3+} + HSO_5^- \rightarrow Co^{2+} + SO_5\bullet^- + H^+ \tag{25}$$

$$Co^{2+} + SO_4\bullet^- \rightarrow Co^{3+} + SO_4^{2-} \tag{26}$$

$$SO_4\bullet^- + OH^- \rightarrow SO_4^{2-} + HO\bullet \tag{27}$$

$$SO_4\bullet^- \,(SO_5\bullet^-, HO\bullet) + organics \rightarrow \ldots\ldots \rightarrow CO_2 + H_2O \tag{28}$$

Therefore, three types of reactive radicals including sulphate ($SO_4\bullet^-$), peroxy-sulphate ($SO_5\bullet^-$) and hydroxyl radicals $HO\bullet$ can be generated during PMS activation by cobalt, although peroxy-sulphate radical is less efficient to attack the organic compounds due to its weak oxidizing ability ($E(SO_5\bullet^- / SO_4^{2-}) = 1.1$ V).

Transition mixed metal spinels, iron-based heterogeneous catalysts and other transition metal oxides have also been studied in this system [89]. However, the use of perovskites for activation of PMS has not been revised. In the present review we show some examples of the application of oxide-like perovskites as activators of PMS for the degradation of organics in waters (Table 2).

Table 2. Degradation of pollutants by the combination of peroxymonosulfate (PMS) and oxides type perovskite.

Reference	Catalyst	Target Pollutant	Concentration Catalyst	Concentration Pollutant	PMS	Treatment Efficiency Degradation	Treatment Efficiency TOC Removal
[95]	$ACoO_3$ (A = La, Ba, Sr and Ce)	phenol	0.2 g/L	20 mg/L	0.1 mM	95% in 180 min ($LaCoO_3$ and $SrCoO_3$) 80% in 180 min ($BaCoO_3$ and $CeCoO_3$)	81% in 6 h ($LaCoO_3$ and $SrCoO_3$) 35% in 6 h ($BaCoO_3$ and $CeCoO_3$)
[96]	$PrBaCo_2O_{5+\delta}$	phenol, MB	0.1 g/L (for phenol) 0.05 g/L (for MB)	20 mg/L phenol 10 mg/L MB	21.2 mM (for phenol) 2.3 mM (for MB)	100% phenol in 30 min 100% MB in 15 min	39.3% at pH 2, 82% at pH 9
[97]	$LaCoO_3/ZrO_2$	RhB	0.1 g/L	10 mg/L	0.1 g/L	100% in 60 min	-
[98]	$SrCo_{1-x}Ti_xO_{3-\delta}$ (SCT_x, x = 0.1, 0.2, 0.4, 0.6)	phenol	0.1 g/L	20 mg/L	11.9 mM	85% in 15 min for $SCT_{0.4}$	76.2% in 2 h
[99]	$Ba_{0.5}Sr_{0.5}Co_{0.8}Fe_{0.2}O_{3-\delta}$ (BSCF)	phenol	0.1 g/L	20 mg/L	6.5 mM	100% in 30 min	-
[100]	$LaCo_{0.4}Cu_{0.6}O_3$	phenol	0.1 g/L	20 mg/L	0.65 mM	100% in 12 min	-
[101]	$LaCo_{1-x}Mn_xO_{3+\delta}$ (LCM, x = 0, 0.3, 0.5, 0.7 and 1.0)	phenol	0.1 g/L	20 mg/L	3.25 mM	100% in 20 min for $LaCoO_{3.002}$ 100% in 40 min for $LaCo_{0.5}Mn_{0.5}O_{3.053}$	67% in 40 min
[102]	$LaCoO_3$ (LCO) LCO-SiO_2, CTAB-LCO	PBSA	0.5 g/L	5 mg/L	5 mM	100% in 30 min (LCO-SiO_2) 100% in 5 min (LCO and CTAB-LCO)	-
[103]	$LaCoO_3$	four herbicides (metazachlor, tembotrione, tritosufuron and ethofumesate)	0.5 g/L	1 mg/L (each)	0.1 mM	95% metazachlor, 85% tembotrione, 5% tritosufuron; 45% ethofumesate t = 60 min; pH = 7	-
[104]	$LaMO_3$ (M = Fe, Ni, Cu, Co)	RhB	0.1 g/L	10 mg/L	0.6 mM	42% in 60 min ($LaFeO_3$) 60% in 60 min ($LaNiO_3$) 45% in 60 min ($CuFeO_3$) 98% in 60 min ($LaFeO_3$)	-
[105]	$LaFeO_3$	diclofenac	0.6 g/L	0.15 mM	0.3 mM	100% in 1 h	50% in 2 h

Table 2. *Cont.*

Reference	Catalyst	Target Pollutant	Concentration		Treatment Efficiency	
			Catalyst	Pollutant	PMS	Degradation	TOC Removal
[107]	BiFeO$_3$ + visible light	RhB	1 g/L	5 mg/L	5 mM	63% in 40 min (25 °C) 93% in 40 min (45 °C)	-
[108]	LaCoO$_3$-TiO$_2$ (Co:Ti = 0:1–1:0) + UVA light	four herbicides: metazachlor, tembotrione, tritosufuron and ethofumesate	0.5 g/L	1 mg/L (each)	0.1 mM	90% metazachlor, 97% tembotrione, 20% tritosufuron; 70% ethofumesate t = 60 min; pH = 7	-
[109]	Sr$_2$CoFeO$_6$ + UV light	bisphenol F	0.3 g/L	20 mg/L	0.1 mM	75% in 2 h	90% in 6 h

MB: Methylene Blue; PBSA: 2-phenyl-5-sulfobenzimidazole acid; RhB: Rhodamine B; Rh6G: Rhodamine 6G.

As cobalt is catalogued as one of the best transition metals in the homogeneous activation of PMS, most of the described examples for the heterogeneous activation by perovskites are based on those containing cobalt in B position, alone [95–97] or partially substituted by other cations [98–101]. And as occurred in the Fenton-like reactions, phenol is again one of the most studied pollutant [95,96,98–101].

A series of cobalt-perovskite catalysts, ACoO$_3$ (A = La, Ba, Sr and Ce) was tested by Hammouda et al. [95] in the degradation of phenol by action of PMS. LaCoO$_3$ and SrCoO$_3$ showed the best catalytic performance, leading to a depletion of 95% of phenol in 3 h and a TOC removal of 65% in 6 h, in contrast to the 80% of removed phenol and 35% of mineralization degree reached with BaCoO$_3$ and CeCoO$_3$. Phenol degradation followed the pseudo first order kinetics and the intermediate formed were identified as catechol, hydroquinone and benzoquinone. Only between 7 and 12% of phenol was removed by physical adsorption. The activity was not related to the textural properties but to the content of cobalt of samples, the removal of phenol increasing with cobalt amount. The degradation of phenol by PMS in absence of perovskite was only of 10% after 3 h, which indicated the low oxidation power of PMS as compared to sulphate radicals formed in the presence of catalyst.

Su et al. [96] found that both hydroxyl and sulphate radicals were responsible for the degradation of phenol and methylene blue (MB) in the presence of PMS and a mixed ionic–electronic conducting (MIEC) double perovskite, PrBaCo$_2$O$_{5+\delta}$ (PBC) over a wide pH range, although the sulphate were the major radicals for promoting the degradation of organics. In addition, the oxygen vacancies in perovskite structure played a key role in the activation of PMS and in facilitating easier valence-state changes of the cobalt ions. The PBC catalysed the phenol oxidation with a TOF that was ~196-fold higher than that of the classical Co$_3$O$_4$ spinel and 100% phenol was removed in 30 min. In the case of MB only 15 min were necessary to produce the complete degradation.

Zirconia-supported LaCoO$_3$ perovskite, LaCoO$_3$/ZrO$_2$ and its corresponding LaCoO$_3$ powder, were used to degrade RhB in the presence of PMS [97]. The nanocomposite showed a much higher catalytic activity than LaCoO$_3$ to activate PMS, in spite of the fact that it contained only 12.5 wt% of LaCoO$_3$. RhB was completely degraded in only 60 min and the nanocomposite could be reused for several cycles without activity loss.

Different perovskites of cobalt in B position partially substituted by Ti, Fe, Cu or Mn have been tested in the degradation of phenol [98–101]. In this sense, SrCo$_{1-x}$Ti$_x$O$_{3-\delta}$ (SCTx, x = 0.1, 0.2, 0.4, 0.6) perovskites exhibited an excellent activity for phenol degradation under a wide pH range, leading to a faster oxidation than Co$_3$O$_4$ and TiO$_2$ [98]. The order of activity was SCT0.2 ≈ SCT0.1 > SCT0.4 > SCT0.6, therefore the rate of phenol oxidation decreasing with the content of cobalt. The effects of operating conditions and initial pH on the catalytic activity were studied for the SCT0.4/PMS system. At pH ≥ 7 the catalyst led to an optimized performance in terms of higher TOC removal, minimum Co leaching and good catalytic stability, which can overcome the common problems of Fenton reaction and provide a promising application for real wastewater treatments under neutral or alkaline conditions. Less than 5% of phenol was removed by adsorption during the 90 min period and the same amount was degraded by PMS in 90 min in the absence of catalyst.

Ba$_{0.5}$Sr$_{0.5}$Co$_{0.8}$Fe$_{0.2}$O$_{3-\delta}$ (BSCF) perovskite was very effective for PMS activation to produce free radicals and the subsequent degradation of phenol [99]. On the contrary, it was not active in the production of radicals from activation of other peroxides, such as H$_2$O$_2$ or peroxydisulfate (PDS). Authors found that the oxygen vacancies and the metal ions in A position with a less electronegativity than cobalt in the perovskite structure play a key role by conferring cobalt sites a high charge density for interacting with PMS via a rapid charge transfer process and to produce free radicals, resulting in a higher activity when compared to a Co$_3$O$_4$ spinel. Thus, 100% of phenol was removed in 30 min with BSCF, in contrast to the 45% of degradation reached with Co$_3$O$_4$. Authors concluded that the PMS activation by BSCF gave rise to the generation of both hydroxyl and sulphate radicals.

LaCo$_{1-x}$Cu$_x$O$_3$ (x = 0–1) perovskites prepared via sol-gel method with citric acid as organic complexing agent were also tested in the PMS-phenol system [100]. LaCo$_{0.4}$Cu$_{0.6}$O$_3$ was the best catalyst, showing a removal efficiency of 100% in only 12 min and a TOF value of 1 h^{-1}, which was 2.5 times higher than that obtained by Duan et al. (0.4 h^{-1}) [99] for the same concentrations of catalyst and pollutant, although the PMS dosage used in the first case was ten times lower. No significant change on surface of the catalyst was observed after the oxidation reaction, proving the high stability of LaCo$_{0.4}$Cu$_{0.6}$O$_3$, although an activity loss of 20% was produced after fourth cycle, due to the poisoning of active sites by adsorption of degradation intermediates. The redox species involved in the mechanism were not only the Co^{2+}/Co^{3+} pair but also the Cu$^+$/Cu^{2+} couple, which reacted with both SO$_4\bullet^-$ and HO$\bullet$.

Miao 2018 [101] synthesized a series of LaCo$_{1-x}$Mn$_x$O$_{3+\delta}$ (LCM, x = 0, 0.3, 0.5, 0.7 and 1.0) perovskites, calcined at different temperatures, showing over stoichiometric oxygen. Authors found that the interstitial oxygen plays a key role in the catalytic activity for degradation of phenol, in such way that a proper amount of interstitial oxygen promotes the electron transfer rate of the perovskite but an excess hinders this process. The most active catalyst was that containing only cobalt in B position, that is, LaCoO$_3$, which led to a complete depletion of phenol in only 20 min. Among all the substituted catalysts, LaCo$_{0.5}$Mn$_{0.5}$O$_{3.053}$, calcined at 900 °C, exhibited the best performance, due to its high interstitial oxygen ion diffusion rate. Furthermore, its stronger relative acidity contributed to an enhanced stability. Phenol was completely degraded with this catalyst after 40 min. Considering that manganese ions are much cheaper and less toxic than cobalt ions, these substituted perovskites are an appropriate alternative for the activation of PMS.

LaCoO$_3$ perovskite has proved to be very efficient for the activation of PMS in the degradation of different organic pollutants [102–104]. The degradations of aqueous solutions of 2-phenyl-5-sulfobenzimidazole acid (PBSA) using PMS activated with LaCoO$_3$ perovskites prepared by three different methods was investigated by Pang et al. [102]. LaCoO$_3$ was prepared by a normal precipitate method (sample named as LCO), by introduction of cetyltrimethylammonium bromide (CTAB-LCO) and by a hydrothermal method with the adding of silicon (LCO-SiO$_2$). LCO-SiO$_2$ was active in a wider pH range (4–8), leading to a complete removal of PBSA in 30 min and showing a very low leaching of metal ions. On the contrary, LCO and CTAB-LCO presented a contribution of the homogeneous reaction to the total activity, due to the leached metals, which resulted in the PBSA depletion of 100% in only 5 min. From studies with radical quenchers and from identified intermediates authors concluded that for LCO-SiO$_2$ the activation of PMS resulted from the combination of SO$_4\bullet^-$ and electronic transfer reaction. However, in the case of LCO and CTAB-LCO, both SO$_4\bullet^-$ and HO$\bullet$ radicals were involved.

More recently, Solís et al. [103] have reported the combination of LaCoO$_3$ and PMS for the removal of various aqueous herbicides (metazachlor, tembotrione, tritosufuron and ethofumesate). The catalyst amount exerted a positive influence on herbicides conversion, which increased when the load of catalyst did from 0.5 to 1.5 g/L. An increment in the pH values reduced cobalt leaching and decreased PMS depletion. As the point of zero charge (PZC) value of LaCoO$_3$ was 9.08, at acidic or neutral pH the catalyst surface is positively charged and as a result it interacts more easily with the anions from PMS. With respect to the influence of PMS concentrations, those equal or above 0.5 mM produced the instantaneous removal of metazachlor, tembotrione and ethofumesate while tritosulfuron required almost one hour to be completely degraded when 0.5 mM of PMS was used.

The high activity of cobalt ions for PMS activation was confirmed by Lin et al. [104], who tested a series of LaMO$_3$ perovskites (M=Co, Cu, Fe and Ni) in the removal of RhB. Once more, LaCoO$_3$ was the most active, followed by LaNiO$_3$, LaCuO$_3$ and LaFeO$_3$. The mechanism of reaction was studied by addition of different scavengers or radical inhibitors. Authors found that both Co^{3+}/Co^{2+} and La^{3+}/La^{4+} ions decomposed PMS yielding mainly sulphate radicals and hydroxyl radicals in lesser extent. By comparison of the obtained rate constants with other from literature, authors concluded that LaNiO$_3$, LaCuO$_3$ and LaFeO$_3$ were no competitive with other existing catalysts, because of the low activity of Ni, Cu and Fe for PMS activation, in contrast to LaCoO$_3$, which exhibited a rate constant comparable or even higher than those reported for other catalysts.

Iron in B position of perovskites has resulted to be also an efficient cation for the PMS activation [105,106]. The oxidative degradation of diclofenac (DCF), a non-steroidal anti-inflammatory drug, was carried out in the presence of $LaFeO_3$ and PMS [105]. DFT studies allowed authors to conclude that a strong interaction occurs between the Fe (III) sites on $LaFeO_3$ surface and PMS, with the formation of an inner-sphere complex and the transfer of electrons from PMS to Fe (III). Sulphate radicals were identified as the major responsible for DCF degradation by the $LaFeO_3$/PMS system. Although 100% of DCF was removed in only 1 h, the mineralization was only of 50% and fifteen different intermediates were formed.

$La_{0.8}Ca_{0.2}Fe_{0.94}O_{3-\delta}$ and $Ag-La_{0.8}Ca_{0.2}Fe_{0.94}O_{3-\delta}$ were tested by Chu et al. [106] in the removal of phenol, MB and rhodamine 6G. From electrochemical impedance spectra, authors concluded that Ag nanoparticles and lattice oxygen vacancies improve the p-type conductivity of the perovskite. Furthermore, the O_2 of solution is adsorbed on the oxygen vacancies and as a consequence, in order to replace the lost oxygen, more $SO_5\bullet^-$ react generating more sulphate radicals. Under the optimum conditions, around 84–90% of MB was degraded in only 45 min. When using $Ag-La_{0.8}Ca_{0.2}Fe_{0.94}O_{3-\delta}$ rhodamine 6G and phenol were completely removed in 15 and 10 min, respectively. This perovskite was also very efficient for the removal of *Escherichia coli*.

4.2. Peroxymonosulfate Activated by the Combination of Perovskites and Light Irradiation (PMS/Catalyst/Light)

The combination of PMS activation by a perovskite and light irradiation has been applied for the degradation of different pollutants [107–109]. Rhodamine B (RhB) was used as model of organic pollutant for studying the PMS activation by a $BiFeO_3$ microsphere in presence of visible light [107]. To confirm the contribution of the oxidizing radical species, ethanol (EtOH), t-butanol (t-BuOH) and 1,4-benzoquinone (BQ) were employed as radical scavengers. The results indicate that the main generated radical species during the activation of PMS by $BiFeO_3$ were $HO\bullet$ and $SO_4\bullet^-$ radicals. However, the $O_2\bullet^-$ radicals, which are formed at longer reaction times, also play an important role in the degradation of RhB. Authors explained the reaction mechanism as follows. $BiFeO_3$, with narrow band gap energy (1.92 eV), can be easily excited by visible light and the electrons and holes generated by action of light can react with Fe^{3+} and RhB, respectively. Simultaneously, Fe^{3+} and Fe^{2+} can also activate PMS to yield $SO_4\bullet^-$ and $SO_5\bullet^-$ radicals, which degrades RhB. About 63% of RhB was degraded in 40 min by $BiFeO_3$/PMS/vis light system, in contrast to 43% of removal in the absence of perovskite. The catalyst was used in three consecutive cycle runs without significant loss of photocatalytic activity.

The combination of a cobalt perovskite ($LaCoO_3$) with a photocatalyst (TiO_2) in different molar ratios (Co:Ti = 0:1–1:0) was used to activate PMS for the oxidation of a mixture of four herbicides (metazachlor, tembotrione, tritosulfuron and ethofumesate) [108]. Same authors had previously tested $LaCoO_3$/PMS system for the degradation of the same herbicides [103]. In general, $LaCoO_3$-TiO_2 with Co:Ti ratios in the range 0.1:1 to 0.5:1 showed a higher activity than the rest of solids tested, although a Co/Ti ratio of 0.1:1 was enough to reach enhanced degradation rates when compared to pristine titania or pure perovskite. The number of degraded herbicides by $LaCoO_3$-TiO_2/PMS system was 3.5–5 times higher in the presence of UVA-light and the reaction followed a second order kinetic, depending of concentration of both PMS and herbicides. As complete mineralization was not achieved, authors carried out studies to assess the potential phytotoxicity of the accumulated intermediates, concluding that all samples did show no phytotoxicity after 180 min of treatment for PMS concentrations ≥ 0.15 mM.

A double cobalt perovskite, Sr_2CoFeO_6, was tested in the mineralization of bisphenol F (BPF) in neutral medium by activation of PMS under UV irradiation [109]. Neither direct UV photolysis nor PMS alone degraded the BPF and Sr_2CoFeO_6 exhibited higher activity (75% of BPF degradation in 2 h) than the corresponding single perovskites, $SrCoO_3$ (60%) and $SrFeO_3$ (35%), caused probably by an accelerated reduction of Fe^{3+} in the presence of cobalt ions. UV irradiation also improved the mineralization degree of BPF and values of TOC removal in 6 h increased from 65% to 90% when Sr_2CoFeO_6/PMS system was irradiated. As a novel aspect of this work, authors studied the influence of chemicals co-existing in the natural water matrix, such as humic acid and inorganic anions (Cl^-, HCO_3^- and CO_3^{2-}), on the inhibition of degradation of bisphenol.

5. Photocatalytic Oxidation (Light/catalyst)

Within AOPs, photocatalyst-based degradation methods represent an interesting research field where there has been continuous development. Heterogeneous photocatalysis is widely recognized as an effective technology for treating waters containing some refractory organic compounds through the photogeneration of oxidizing radicals such as HO• and O^{2-}•. It is a green technology with broader application prospect and compared with the traditional chemical oxidation, photocatalysis is usually non-toxic, non-corrosive and harmless to the environment. The photocatalytic oxidation is based on the use of a semiconductor and ultraviolet-visible (UV-vis) radiation (see Figure 4). The fundamental step of the process is generation of electron-hole pairs, which requires absorption of photons with adequate energy and promotion of electrons from the valence band to the conduction band. The photogenerated charge carriers participate in a series of reactions producing highly reactive radicals. One of the most relevant applications of this technique is the degradation of environmental pollutants in aqueous wastewater into less harmful products. These treatments are very appropriate because of their on-place use and because they do not have extra energy consumption. The degradation of organics is normally accomplished of semiconductors such as zinc oxide (ZnO) or titanium dioxide (TiO_2). The latter is currently the most popular photocatalyst due mainly to its specific photocatalytic properties like for example strong oxidizing power, high chemical stability and relative inexpensiveness. Recently, different photocatalytic materials capable of efficiently working with sunlight, based mostly on TiO_2 and their combination with solar collectors, have been revised [110].

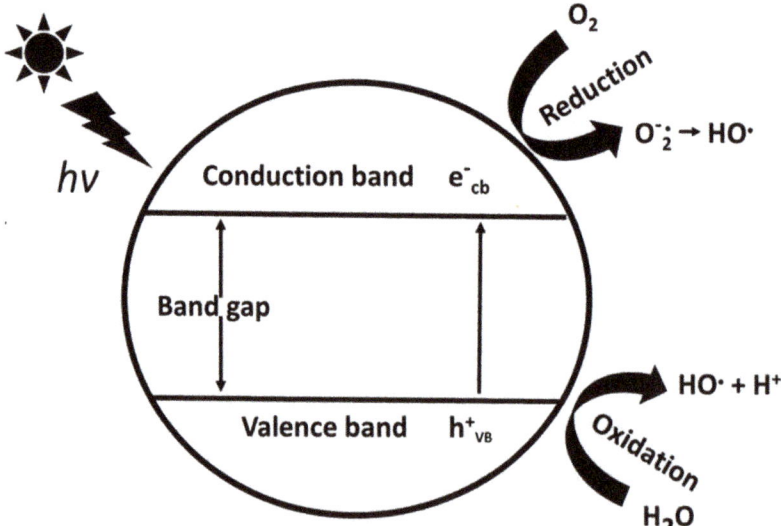

Figure 4. Schematic illustration of a model photocatalytic system showing the contribution of hole-electron couples to the formation of radicals.

Unfortunately, the use of TiO_2 catalysts is limited by the large band gap of TiO_2 (Eg = 3.2 eV), being active only under UV light. Considering that the sunlight is composed mainly of visible light (43%) while only 4% of the spectrum is UV light, visible-light photocatalytic performance is desirable in order to effectively utilize the sunlight. In this sense, considerable effort is being devoted to developing alternative heterogeneous photocatalysts, which are active under visible light. Among the various materials, some perovskites have been considered as a promising photocatalysts, since they present high activity in the long band of visible-light. They can be used alone or combined with TiO_2 in form of composites, with the aim of narrowing the bandgap of this oxide.

One of the perovskites more widely studied as photocatalyst is $LaFeO_3$ [111–122], due to its narrow band gap (often less than 3.0 eV), which can be excited easily under visible light or UV light irradiation. It can be used auto supported or in form of composites and rhodamine B (RhB) has been tested by different authors as model molecule in the evaluation of its photocatalytic activity [111,112,114,119–122].

In Reference [111] $LaFeO_3$ particles prepared by sol-gel method were able to degrade RhB in 24% under visible irradiation, showing a higher activity than that of international $P-25TiO_2$. Authors found a reverse correlation between crystallite size and photocatalytic activity, the most active sample being that exhibiting the smallest size. Values of degradation comprised between 15 and 50% were reached when $LaFeO_3$ nanoparticles were prepared by using silica SBA-16 as template [112]. The high surface area and crystallinity of samples were responsible for the adsorption and photocatalytic degradation of RhB, respectively.

Li et al. [114] prepared different samples containing $LnFeO_3$ (Ln = La, Sm) nanoparticles by sol–gel method at different calcination temperatures. 80% of RhB was degraded in 2 h by $LaFeO_3$ under visible light, in contrast to the 20% obtained with $SmFeO_3$. When H_2O_2 was added to the reaction media, the photocatalytic activity improved due to the synergistic effect between the semiconductor photocatalysis and Fenton-like reaction. Complete degradation of RhB was achieved after 3 h of reaction when microspheres composed of perovskite $LaFeO_3$ nanoparticles, prepared by hydrothermal method, were used as photocatalysts [119]. In this case, the hydrothermal reaction conditions and the concentration of citric acid played an essential role in the development of $LaFeO_3$ microspheres. The efficiency of $LaFeO_3$ microspheres was higher than that of $LaFeO_3$ prepared by microwave assisted method [120] (95% of degradation of RhB in 3 h).

As an alternative way to obtain perovskite-type nanoparticles, graphitic carbon nitride (g-C_3N_4) has been combined with $LaFeO_3$ by using a solvothermal method [121] according to the scheme of preparation shown in Figure 5. The synergistic interaction between $LaFeO_3$ and g-C_3N_4 improved the separation efficiency of photogenerated electron-hole pairs. Then the photocatalytic activity for the degradation of RhB was 19 times higher than that of $LaFeO_3$ under visible light irradiation. In addition, the catalysts kept excellent stability after four cycles.

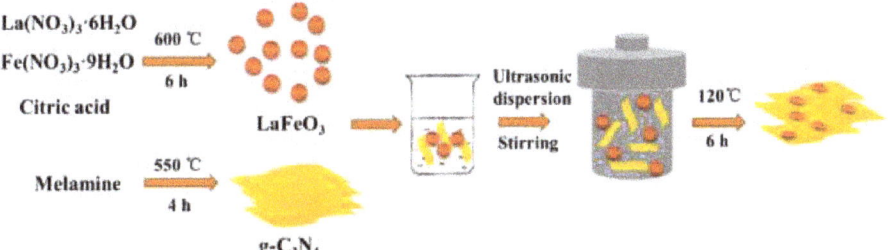

Figure 5. Schematic illustrating the synthesis of the $LaFeO_3$/g-C_3N_4 heterojunction. With permission from [121].

To achieve more homogeneous $LaFeO_3$ nanoparticles, Ren et al. [122] used reduced graphene oxide as template. They obtained nanoparticles anchored on graphene oxide by combining the sol-gel method and high-temperature annealing. The $LaFeO_3$–rGO can work under visible-light irradiation as an efficient catalyst for the degradation of RhB and MB, the bandgap of $LaFeO_3$ nanoparticles on reduced graphene oxide being of 1.86 eV. The oxidation process was dominated by the electron transfer since the presence of rGO facilitated the electron-hole separation.

Methylene blue (MB) has been degraded under the photocatalytic reactions with LaFeO$_3$ perovskites [116,117] and LaFeO$_3$ doped in A position [115,118], whose activity is strongly influenced by the process of synthesis. There are many methods to prepare LaFeO$_3$ such as sol-gel, co-precipitation, electro-spinning, citric acid complex, stearic acid solution combustion or glycine combustion at high temperature. Among these methods, the sol-gel process has been proven to be one of the most effective [116] and MB and methyl orange were completely degraded after visible light irradiation for 4 h of LaFeO$_3$ synthesized by sol-gel method and calcined under vacuum microwave. The photodegradation process of MB on LaFeO$_3$ followed a pseudo-first-order kinetic process.

However, sol-gel method and solid-state reactions need annealing at a high temperature which results in short homogeneity and high porosity of the samples without control on the particle size. As an alternative way, the microwave assisted synthesis allows preparing nanoparticles with small size, narrow size distribution and high reactive ability in a time-saving process. In this regard, LaFeO$_3$ with high crystallinity and sphere-like shape was able to decolorize a MB aqueous solution in only 90 min of exposure to visible light [117].

A strategy to improve the photocatalytic efficiency of perovskites is by doping in A position [115,118]. Thus, a series of LaFeO$_3$ perovskites doped with Li, La$_{1-x}$Li$_x$FeO$_3$ (with x = 0, 3, 5 and 7%) were active under light irradiation for the degradation of MB and arcylon effluents, La$_{0.97}$Li$_{0.03}$FeO$_3$ showing the highest activity [115]. LaFeO$_3$ and Ca-doped LaFeO$_3$, synthesized via reverse microemulsion without additional high temperature calcination process [118] were active in the photocatalytic degradation of MB under the action of visible light. When La$_{0.9}$Ca$_{0.1}$FeO$_3$ was used instead of LaFeO$_3$, the degradation rate of MB improved 30%.

LaFeO$_3$ and its corresponding double perovskite, La$_2$FeTiO$_6$, have also been tested in the degradation of p-chlorophenol [113] under visible light. Authors correlated the photocatalytic activity with differences in structure or surfaces properties. Thus, the optical property in the visible light region and the inferior symmetry at the Fe nucleus of La$_2$FeTiO$_6$ resulted in a better performance with respect to that of LaFeO$_3$.

Apart from LaFeO$_3$ other perovskites have been used as photocatalysts. In this regard, the fact that LaCoO$_3$ hollow nanospheres exhibit a band gap of 2.07 eV makes this compound a promising candidate material for photocatalytic applications. Fu et al. [123] studied the photocatalytic degradation of MB, methyl orange and neutral red under UV irradiation. UV–vis analysis showed that LaCoO$_3$ hollow nanospheres exhibited excellent photocatalytic activity, reaching degradation values near 90% in 100 min for the three contaminants. Moreover, the authors discussed the influence of temperature and time of calcination on the structures of LaCoO$_3$ and its formation mechanism.

Alkali earth titanates, such as SrTiO$_3$, are perovskite-type oxides based on a Ti-O polyhedron, showing a similar energy band structure to that of TiO$_2$. Furthermore, SrTiO$_3$ shows a wide absorption band in the ultraviolet region in the 300–400 nm range. As a result it can be used as photocatalyst, whose activity can be improved by incorporation of CeO$_2$ to the structure, which shows strong UV absorption in the 300–450 nm range. In this sense, a SrTiO$_3$/CeO$_2$ composite was used in the photodegradation of two azo dyes, C.I. Reactive Black 5 [124] and C.I. Direct Red 23 [125]. The influence of pH on the photoactivity was studied in the first case. Authors extended the study to the influence of other parameters, such as catalyst dose, concentration of dye, pH value, irradiation intensity and use of KI as scavengers, in the second one. Under the optimum conditions, a complete degradation of C.I. reactive black 5 and C.I. direct red 23 was achieved in 120 min and 60 min, respectively. Authors proposed a tentative degradation pathway based on the sensitization mechanism of photocatalysis.

Recently, bismuth-based perovskites have also attracted interest as photocatalysts due to their particular electronic structure that reduces the charge mobility and the band gap to ~2 eV [126–129]. BaBiO$_3$ powders with a base centred monoclinic structure exhibited good activity for the water-splitting reaction and the degradation of rhodamine B dye under visible light [126]. Authors demonstrated that the catalytic activity strongly depended on the crystallinity of the materials, BaBiO$_3$ prepared by solid state being the catalyst with the highest crystallinity, the lowest resistance to the charge transfer and the greatest photocatalytic performance. Another bismuthate, KBiO$_3$, was investigated as a visible-light-driven photocatalyst in the degradation of organic pollutants, such as RhB, crystal violet, MB and phenol [127]. The difference between the degradation mechanisms of these organic pollutants under the action of KBiO$_3$ depended on competition of the photocatalysis, redox reaction and adsorption mechanisms. In the case of RhB and crystal violet, the redox potentials are higher than that of KBiO$_3$ (1.59 eV) but lower than its band gap energy (2.04 eV), thus only the adsorption and photooxidation controlled the reactions. MB presents a redox potential lower than the band gap of KBiO$_3$ and in this case the reaction was controlled by both the photooxidation and chemical oxidation. In the case of phenol, photooxidation was observed at the end steps of the process whereas quick adsorption and chemical oxidation were determined at the initial stage. A double Bi-perovskite, Bi$_2$Fe$_4$O$_9$, composed of nanoplates, which behaves as a multiband semiconductor [128] was tested in the photocatalytic oxidation to aqueous ammonia oxidation and phenol under visible light irradiation. The catalyst displayed a higher activity when compared to its bulk material. The improvement of photocatalytic performance of Bi$_2$Fe$_4$O$_9$ could be due to the efficient electron-hole separation that may act as electron-hole recombination centres. The photocatalytic performance for phenol oxidation enhanced when an appropriate amount of, H$_2$O$_2$, was added, which can act as a strong electron scavenger and also as a promoter of the Fenton-like reaction.

LaMnO$_3$ perovskite is another promising photocatalyst owing to its catalytic and electrical properties, price, nontoxicity and high stability. However, a high combination rate of electron/hole pairs and the agglomeration of particles are some of the perovskite limitations. Semiconductor coupling with carbon materials is believed to induce cooperative o synergistic interactions retarding the fast recombination of the charge carriers and getting better the photocatalytic activity. In this sense, Huang et al. [130] observed a higher efficiency for photodegradation of acid red C-3GN over a series of LaMnO$_3$-diamond composites than for LaMnO$_3$. In the composites, the perovskite particles are uniformly distributed on the diamond surface creating a network structure, which increases the active sites and the absorption of dye molecules. The composite showed the best photocatalytic activity when the mass ratio was 1LaMnO$_3$/2diamond.

AuNP/KNbO$_3$ have shown photocatalytic activity in the photooxidation of sec-phenethyl alcohol to acetophenone under the action of visible light in the presence of H$_2$O$_2$ [131]. Photophysical properties of KNbO$_3$ and TiO$_2$ are fundamentally similar, with band gaps near 3.2 eV; however, the particle size of KNbO$_3$ presents advantages over TiO$_2$ since small particle size of TiO$_2$ make difficult its separation from reaction solutions. The activity of this AuNP-decorated KNbO$_3$ was superior to that of undecorated KNbO$_3$.

In recent years most researchers have concentrated their attention on modifying perovskite by doping with a transition metal or non-metal to improve the catalytic activity. The doping allows the recombination of centres of electron-hole pairs in the semiconductor particles. For instance, the doping with C and S atoms improved the photocatalytic activity of SrTiO$_3$ for oxidation of 2-propanol [132], because a new absorption edge in the visible light region was produced.

Several articles report the benefit of using nanostructures to improve the photocatalytic activity with respect to the bulk samples [129,133,134]. Thus, perovskite-type BiFeO$_3$ nanoparticles showed increased degradation ability of methyl orange under visible light irradiation with respect to bulk BiFeO$_3$, probably due to the higher surface area of the nanoparticles [129]. As a result, more than 90% of MO was decolorized after 8 h under UV-vis irradiation and after 16 h under visible light when BiFeO$_3$ nanoparticles were used. In contrast, only 70% of dye was degraded after 16 h by the action of UV-vis light in the presence of bulk BiFeO$_3$. Dong et al. [133] found that LaCoO$_3$ nanofibers prepared at different temperatures presented better photocatalytic activity for the degradation of RhB than LaCoO$_3$ particles. Nanofibers exhibited an increased surface, providing more photocatalytic active sites on the inner/outer surfaces that led to a complete degradation of RhB in only 50 min for the best catalyst, which resulted to be that with a high degree of crystallinity of LaCoO$_3$ nanofibers and containing some residual carbon. PrFeO$_3$ porous nanotubes showed high optical absorption in the UV-visible region and an energy band gap of 1.97 eV [134], displaying a higher photocatalytic activity in the degradation of RhB than PrFeO$_3$ nanofibers or PrFeO$_3$ nanoparticles. Thus, 46.5% RhB was degraded by the nanotube sample in 6 h, whereas decolorization efficiency was 29.5% and 15.7% for nanofibers and nanoparticle samples, respectively. PrFeO$_3$ nanotubes were used for three cycles of photodegradation, showing a slight deactivation.

Silver orthophosphate (Ag$_3$PO$_4$) has been extensively studied due to its good activity as photocatalyst for organic pollutants degradation under the action of visible light [135]. However, the photocorrosion together with the formation of metallic Ag in the surface of catalyst limit the stability and cause a decrease in the activity and reusability. Several strategies have been proposed to improve the photostability of Ag$_3$PO$_4$, among them, the synthesis of composites perovskite-type (ABO$_3$) has attracted considerable attention due to its high catalytic activity. Guo et al. [136] synthesized and characterized several Ag$_3$PO$_4$/LaCoO$_3$ composites with different ratios, which were evaluated in the photocatalytic degradation of bisphenol A. For the composite containing 10% of LaCoO$_3$, the contaminant was completely degrader after 40 min, achieving a TOC removal of 77.3%. Furthermore, the authors investigated in detail the degradation intermediates and the photocatalytic mechanism.

It has been also found that by introducing perovskites in a TiO$_2$ matrix, a beneficial effect in the photocatalytic activity of TiO$_2$ can be obtained. Thus, Gao et al. [137] synthesized multi-modal TiO$_2$-LaFeO$_3$ composite films by a two-step method, which exhibited high photocatalytic activity in the degradation of MB aqueous solution (60% of degradation in 1 h). In comparison with TiO$_2$ and LaFeO$_3$ materials, the composite obtained exhibited good microstructural properties and high specific surface area. The introduction of LaFeO$_3$ not only improved the photocatalytic activity and the hydrophilicity but also influenced the interfacial charge transfer process. The same composite, TiO$_2$-LaFeO$_3$, was prepared by Dhinesh et al. [138] by a hydrothermal method and tested in the degradation of methyl orange in aqueous solution under visible light irradiations. TiO$_2$-LaFeO$_3$ composite exhibited enhanced visible light photocatalytic properties in comparison with LaFeO$_3$ nanoparticles due to a synergetic effect.TiO$_2$ causes the inhibition of the recombination between photoinduced electron and hole pairs and LaFeO$_3$ perovskite played an important role in extending light absorption into the visible region. Halide perovskite CsPbBr$_3$/TiO$_2$ composites [139] also showed an enhanced activity in the selective oxidation of benzyl alcohol to benzaldehyde under visible light irradiation. Action spectra and electron spin resonance studies showed that photo-excited electrons generated within CsPbBr$_3$ were transferred to the conduction band of TiO$_2$, forming, via the reduction of oxygen, superoxide radicals. 50% of benzyl alcohol was oxidized after 20 h of irradiation.

Table 3 summarizes the applications of perovskite-like oxides synthesized by different methods as photocatalysts for the degradation of different organics.

Table 3. Oxides type perovskite applied as photocatalysts for the degradation of different organic contaminants in aqueous solution.

Reference	Catalyst	Preparation Procedure	Contaminant	Light (Irradiation Source)
[111]	LaFeO$_3$	Sol-gel	RhB	Visible
[112]	LaFeO$_3$	Sol-gel, SBA-16 as template	RhB	Visible
[114]	LnFeO$_3$ (Ln = La, Sm)	Sol-gel	RhB	Visible
[119]	LaFeO$_3$	Hydrothermal	RhB	Visible
[120]	ReFeO$_3$ microspheres (Re: La, Sm, Eu, Gd)	Microwave	RhB	Visible
[121]	LaFeO$_3$, g-C$_3$N$_4$	Solvothermal	RhB	Visible
[122]	LaFeO$_3$-rGO	Sol-gel	MB and RhB	Visible
[116]	LaFeO$_3$	Sol-gel, Microwave	Methyl orange and MB	Visible
[117]	LaFeO$_3$	Microwave	MB	Visible
[115]	Li-doped LaFeO$_3$	Sol-gel	MB and arcylon	UV-visible
[118]	LaFeO$_3$, Ca-doped LaFeO$_3$	Reverse microemulsion	MB	Visible
[113]	LaFeO$_3$, La$_2$FeTiO$_6$	Sol-gel	p-Chlorophenol	Visible
[123]	LaCoO$_3$	Surface-ion adsorption	MB, methyl orange and neutral red	UV
[124]	SrTiO$_3$/CeO$_2$	Dry (SrCO$_3$, CeO$_2$)-wet (sol-gel Ti(OC$_4$H$_9$)$_4$) composition	Reactive black 5	UV
[125]	SrTiO$_3$/CeO$_2$	Dry (SrCO$_3$, CeO$_2$)-wet (sol-gel Ti(OC$_4$H$_9$)$_4$) composition	Direct Red 23	UV
[126]	BaBiO$_3$	Solid state, hydrothermal	RhB, water splitting	Visible
[127]	KBiO$_3$	Solid phase heating	RhB, crystal violet, MB, phenol	Visible
[128]	Bi$_2$Fe$_4$O$_9$	Hydrothermal method	phenol, aqueous ammonia	Visible
[130]	LaMnO$_3$–diamond	Sol-gel	Acid red C	Visible
[131]	Au NP/KNbO$_3$	Deposition/precipitation + thermal reduction	Sec-phenethyl alcohol	Visible
[132]	S, C-co-doped SrTiO$_3$	Calcination	2-Propanol	Visible
[129]	BiFeO$_3$ nanoparticles	Sol-gel	Methyl orange	UV-Visible
[133]	LaCoO$_3$ nanofibres	Electrospinning	RhB	UV
[134]	PrFeO$_3$ (nanotubes, nanofibers, nanoparticles)	Electrospinning, annealing	RhB	Visible
[136]	Ag$_3$PO$_4$/LaCoO$_3$	Liquid deposition	Bisphenol	Visible
[137]	LaFeO$_3$/TiO$_2$	Two-step	MB	Fluorescent
[138]	LaFeO$_3$/TiO$_2$	Hydrothermal	Methyl orange	Visible
[139]	CsPbBr$_3$/TiO$_2$	Wet-impregnation	Benzyl Alcohol	Visible

MB: Methylene Blue; RhB: Rhodamine B.

6. Processes under Dark Ambient Conditions

As shown above, to date, great success has been achieved for producing visible-light photocatalysts by doping perovskites or designing composites. Other perovskites have been proven active in AOPS involving an additional chemical oxidant, such as ozone, hydrogen peroxide or peroxymonosulfate. Nevertheless, both kinds of processes using or light or chemical oxidant are expensive, thus limiting their practical applications. Therefore an effort has to be made to find new catalysts capable of working under dark ambient conditions, as potential low-cost alternative for the remediation of waters. A few examples of the application of perovskites for the degradation of organics in waters in dark ambient conditions are shown next.

A layered perovskite La_2NiO_4 crystal [140] can act as a round-the-clock photocatalyst and efficiently degrade phenolic pollutants in the dark. This photocatalyst can produce photoelectrons not only by visible light irradiation but also from some reactant molecules in the dark leading to the degradation of 4-chlorophenol (4-CP). $4\text{-}CP^-$ anions can donate electrons to La_2NiO_4, what is followed by the reaction with dissolved oxygen to generate $O_2\text{-}\bullet$ and reaction with H^+ to form $HO\bullet$ radicals, which can oxidize 4-CP• radicals into CO_2. $LaCoO_{3-x}$ (x= 0-0.075) [141] calcined at different temperatures also displayed activity for the degradation of methyl orange in the dark. 17% of dye was removed with the best catalyst in 45 h in the absence of light. The degradation rate improved under the visible light due to the optical property of $LaCoO_{3-x}$, achieving degradation values of 40%. $Sr_{0.85}Ce_{0.15}FeO_{3-\delta}$ [142] can also work in the dark after thermal activation. $SrFeO_3$ is known as photocatalyst since the bandgap energy values are comprised between 1.80 and 3.75 eV; however, the doping with Ce improves its redox properties and exerts a positive role in oxidation reactions. The catalyst was applied to remove RhB and Orange II from aqueous solution. In the first case the degradation after 7 h only increased from 40 to 60% in the presence of visible light. On the contrary, for the degradation of Orange II it was necessary to irradiate the catalyst, because only 5% was removed in the dark.

Other examples of the use of perovskites as catalysts for degradation of contaminants in aqueous solution in the absence of an oxidant or light have been described. In this regard, methyl orange (MO) was degraded by a layered perovskite, $La_4Ni_3O_{10}$, without additional reagents or external energy [143]. The dye degradation occurred via electrons transfer from the dye molecules to the perovskite and then to the dissolved oxygen, which acted as electrons acceptor. The same dye, MO, was degraded by $LaNiO_{3-\delta}$ under dark ambient conditions (room temperature and atmospheric pressure) [144]. Under the optimum conditions 94.3% of MO was degraded after 4 h. Authors concluded that MO was decomposed by two synergic effects derived from nickel present at the surface of $LaNiO_{3-\delta}$ and the formation of lanthanum carbonate.

$SrFeO_{3-\delta}$ perovskite synthesized by a combined high temperature and high-energy ball milling process was active in the degradation or bisphenol A (BPA) and Acid Orange 8 under dark ambient conditions [145]. The complete degradation of BPA was produced after 24 h, with a TOC removal of 83%. In the case of dye, the full decolorization was attained in only 1 h. By last, more recently, Chen et al. [146] tested a series of $Ca_xSr_{1-x}CuO_{3-\delta}$ (x= 0-1) perovskites in the removal of Orange II dye, widely used in the textile industry. Samples containing a higher amount of Ca were more active for the degradation of Orange II in dark conditions. A depletion in concentration of dye of 80% was reached in only 10 min, which increased to 95% after 1 h. However the mineralization was partial only and some by-products were formed, reaching a TOC removal of 60%. The catalysts were stable after 9 cycles of reusing.

7. Summary and Perspectives

In this paper we have summarized the applications of perovskites as catalysts in heterogeneous advanced oxidation processes for the degradation of pollutants present in waters. Processes have been classified and revised according to the oxidant employed in the process, that is, ozone, hydrogen peroxide and peroxymonosulfate, which can be used alone or combined with light irradiation.

The photocatalytic oxidation, consisting in the activation of a catalyst, in this case, a perovskite, by irradiation with UV or visible light, has also been revised.

The various systems described here using perovskites were shown to effectively degrade and remove specific pollutants from waters. Phenol has been the most studied but other pollutants, such as dyes (especially rhodamine B and methylene blue), phenolic compounds, herbicides and some drugs have also been reported. Although single ozonation (in the absence of catalyst) has been widely used in water and wastewater because is an effective oxidation process, the use of a catalyst improves the decomposition of ozone and the production of hydroxyl radicals and overall increases the degree of mineralization of the contaminants. However, the use of perovskites in this type of processes is very limited and only a few studies have been carried out. Significantly higher is the number of papers related to the application of perovskites in Fenton-like processes, using H_2O_2 as producer of radicals. The stabilization of cations with unusual oxidation states and redox properties in the perovskite network make these oxides good candidates for this type of reaction. Thus, most of reported perovskites contain iron (Fe^{2+}/Fe^{3+}), the Fenton reactant by excellence, in B position but other transition metals, such as copper, manganese or titanium have also been resulted active for the Fenton-like degradation of contaminants. When the H_2O_2/perovskite system is combined with the irradiation of light (photo-Fenton process) the degradation of contaminant and TOC removal generally increase because the rate of production of HO• radicals is higher.

Several examples of perovskites as activators of peroxymonosulfate (PMS) for the production of radicals able to degrade organics present in waters have been presented. It should be remarked that the treatment with PMS is more expensive than other AOPs, due to the price of the reagents. As cobalt is catalogued as one of the best transition metals in the homogeneous activation of PMS, most of the described examples for the heterogeneous activation by perovskites are based on those containing cobalt in B position, alone or substituted by other cations.

It should be pointed that one of the AOPS in which perovskites have been more extensively applied is the heterogeneous photocatalysis, because they present high activity in the long band of visible-light. These processes generally lead to higher TOC removal than processes based on a chemical oxidant. Perovskites have been used in photocatalytic degradation or organics alone or combined with TiO_2 in form of composites, in this way narrowing the bandgap of this oxide. Additionally, the main strategy to improve the photocatalytic activity has been the substitution of the element in A or B position, which leads to the introduction of defects into the narrow band gap and to the formation of oxygen vacancies, which inhibit the recombination between the photogenerated electrons and the holes. Again, as in Fenton-like processes, perovskites containing iron in B position have been the most studied.

Perovskites have resulted been active by alone in the revised AOPs. However, majority of studies were carried out in semi-batch and batch reactors, while continuous fixed bed reactors, which are promising from the practical point of view, have not extensively studied for treatment of real wastewaters. In this sense it can be expected that in the future more studies will be devoted to them.

The low surface area of perovskites implies a limited interaction with the contaminants. In order to increase the surface area, new synthesis methods have been applied and some strategies have been developed, such as their supporting on mesoporous silica supports, honeycombs or the formation of composites. Another strategy to improve catalytic activity of perovskites in AOPs is the hetero-doping in order to produce more active sites of the low-valence B-site transition metals (i.e., Fe^{2+}, Cu^+ and Ti^{3+}) or to introduce oxygen vacancies, which can facilitate the transformation of H_2O_2 (or PMS) into HO•. Furthermore, the use of nanostructures improves the catalytic behaviour with respect to the bulk samples. Considering all of this, we think that research of forthcoming years will be addressed to design new synthesis methods which allow the obtaining of perovskites in form of nanostructures, nanoparticles or nanofibers and also to search new materials based on perovskites containing other different active cations and exhibiting higher surface areas, which can be extended to the removal of other persistent contaminants present in waters.

By last, it should be remarked the high cost of AOPs, involving light or chemical oxidants, which usually have to been constantly fed to keep the process operative. Recent studies reporting promising results of the application of perovskites for degradation of some organics under dark ambient conditions, should encourage researchers in a short-term future to the search of similar systems capable of degrading other contaminants without necessity of using energy or reagents, which would considerably reduce the cost of the process.

Author Contributions: M.L.R.C. chose the topic and designed the organization of paper; M.L.R.C. and E.C. performed the literature search and wrote the article. M.L.R.C. corrected and revised the manuscript according to comments of referees.

Acknowledgments: M.L.R.C. thanks the supporting by the Spanish Ministry of Science and Innovation (CTM2014-56668-R).

Conflicts of Interest: The authors declare no conflict of interest.

References

1. Venkatadri, R.; Peters, R.W. Chemical oxidation technologies: Ultraviolet light/hydrogen peroxide, Fenton's reagent and titanium dioxide-assisted photocatalysis. *Hazard. Waste Hazard. Mater.* **1993**, *10*, 107–149. [CrossRef]
2. Skoumal, M.; Cabot, P.-L.; Centellas, F.; Arias, C.; Rodriguez, R.M.; Garrido, J.A.; Brillas, E. Mineralization of paracetamol by ozonation catalyzed with Fe^{2+}, Cu^{2+} and UVA light. *Appl. Catal. B* **2006**, *66*, 228–240. [CrossRef]
3. Rosenfeldt, E.J.; Chen, P.J.; Kullman, S.; Linden, K.G. Destruction of estrogenic activity in water using UV advanced oxidation. *Sci. Total Environ.* **2007**, *377*, 105–113. [CrossRef] [PubMed]
4. Poyatos, J.M.; Munio, M.M.; Almecija, M.C.; Torres, J.C.; Hontoria, E.; Osorio, F. Advanced oxidation processes for wastewater treatment: State of the art. *Water Air Soil Pollut.* **2010**, *205*, 187–204. [CrossRef]
5. Gogate, P.R.; Pandit, A.B. A review of imperative technologies for wastewater treatment. I: Oxidation technologies at ambient conditions. *Adv. Environ. Res.* **2004**, *8*, 501–551. [CrossRef]
6. Gogate, P.R.; Pandit, A.B. A review of imperative technologies for wastewater treatment. II: Hybrid methods. *Adv. Environ. Res.* **2004**, *8*, 553–597. [CrossRef]
7. Marquez, J.J.R.; Levchuk, I.; Sillanpaa, M. Application of catalytic wet peroxide oxidation for industrial and urban wastewater treatment: A review. *Catalysts* **2018**, *8*, 673/1–673/18.
8. Neyens, E.; Baeyens, J. A review of classic Fenton's peroxidation as an advanced oxidation technique. *J. Hazard. Mater.* **2003**, *98*, 33–50. [CrossRef]
9. Kumar, S.M. Degradation and mineralization of organic contaminants by Fenton and photo-Fenton processes: Review of mechanisms and effects of organic and inorganic additives. *Res. J. Chem. Environ.* **2011**, *15*, 96–112.
10. Oller, I.; Malato, S.; Sanchez-Perez, J.A. Combination of advanced oxidation processes and biological treatments for wastewater decontamination. A review. *Sci. Total Environ.* **2011**, *409*, 4141–4166. [CrossRef] [PubMed]
11. Ferri, D.; Forni, L. Methane combustion on some perovskite-like mixed oxides. *Appl. Catal. B* **1998**, *16*, 119–126. [CrossRef]
12. Zhu, J.; Thomas, A. Perovskite-type mixed oxides as catalytic material for NO removal. *Appl. Catal. B* **2009**, *92*, 225–233. [CrossRef]
13. Tejuca, L.G.; Fierro, J.L.G.; Tascon, J.M.D. Structure and reactivity of perovskite-type oxides. *Adv. Catal.* **1989**, *36*, 237–328.
14. Pena, M.A.; Fierro, J.L.G. Chemical Structures and Performance of Perovskite Oxides. *Chem. Rev.* **2001**, *101*, 1981–2017. [CrossRef] [PubMed]
15. Xu, J.-J.; Xu, D.; Wang, Z.-L.; Wang, H.-G.; Zhang, L.-L.; Zhang, X.-B. Synthesis of perovskite-based porous $La_{0.75}Sr_{0.25}MnO_3$ nanotubes as a highly efficient electrocatalyst for rechargeable lithium-oxygen batteries. *Angew. Chem. Int. Ed* **2013**, *52*, 3887–3890. [CrossRef] [PubMed]
16. Sekhar, P.K.; Mukundan, R.; Brosha, E.; Garzon, F. Effect of perovskite electrode composition on mixed potential sensor response. *Sens. Actuators B* **2013**, *183*, 20–24. [CrossRef]

17. Mori, M.; Itagaki, Y.; Sadaoka, Y. Effect of VOC on ozone detection using semiconducting sensor with SmFe$_{1-x}$Co$_x$O$_3$ perovskite-type oxides. *Sens. Actuators B* **2012**, *163*, 44–50. [CrossRef]
18. Tavakkoli, H.; Yazdanbakhsh, M. Fabrication of two perovskite-type oxide nanoparticles as the new adsorbents in efficient removal of a pesticide from aqueous solutions: Kinetic, thermodynamic and adsorption studies. *Microporous Mesoporous Mater.* **2013**, *176*, 86–94. [CrossRef]
19. Royer, S.; Duprez, D.; Can, F.; Courtois, X.; Batiot-Dupeyrat, C.; Laassiri, S.; Alamdari, H. Perovskites as substitutes of noble metals for heterogeneous catalysis: Dream or reality. *Chem. Rev.* **2014**, *114*, 10292–10368. [CrossRef] [PubMed]
20. Soleymani, M.; Moheb, A.; Babakhani, D. Hydrogen peroxide decomposition over nanosized La$_{1-X}$Ca$_X$MnO$_3$ (0 ≤ X ≤ 0.6) perovskite oxides. *Chem. Eng. Technol.* **2011**, *34*, 49–55. [CrossRef]
21. Ariafard, A.; Aghabozorg, H.R.; Salehirad, F. Hydrogen peroxide decomposition over La$_{0.9}$Sr$_{0.1}$Ni$_{1-x}$Cr$_x$O$_3$ perovskites. *Catal. Commun.* **2003**, *4*, 561–566. [CrossRef]
22. Lee, Y.N.; Lago, R.M.; Fierro, J.L.G.; Gonzalez, J. Hydrogen peroxide decomposition over Ln$_{1-x}$A$_x$MnO$_3$ (Ln = La or Nd and A = K or Sr) perovskites. *Appl. Catal. A* **2001**, *215*, 245–256. [CrossRef]
23. Garrido-Ramirez, E.G.; Theng, B.K.G.; Mora, M.L. Clays and oxide minerals as catalysts and nanocatalysts in Fenton-like reactions—A review. *Appl. Clay Sci.* **2010**, *47*, 182–192. [CrossRef]
24. Xu, L.; Wang, J. Magnetic nanoscaled Fe$_3$O$_4$/CeO$_2$ composite as an efficient Fenton-Like heterogeneous catalyst for degradation of 4-chlorophenol. *Environ. Sci. Technol.* **2012**, *46*, 10145–10153. [CrossRef] [PubMed]
25. Rahim Pouran, S.; Abdul Raman, A.A.; Wan Daud, W.M.A. Review on the application of modified iron oxides as heterogeneous catalysts in Fenton reactions. *J. Clean. Prod.* **2014**, *64*, 24–35. [CrossRef]
26. Pereira, M.C.; Oliveira, L.C.A.; Murad, E. Iron oxide catalysts: Fenton and Fenton-like reactions—A review. *Clay Miner.* **2012**, *47*, 285–302. [CrossRef]
27. Nidheesh, P.V. Heterogeneous Fenton catalysts for the abatement of organic pollutants from aqueous solution: A review. *RSC Adv.* **2015**, *5*, 40552–40577. [CrossRef]
28. Ramirez, J.H.; Costa, C.A.; Madeira, L.M.; Mata, G.; Vicente, M.A.; Rojas-Cervantes, M.L.; López-Peinado, A.J.; Martín-Aranda, R.M. Fenton-like oxidation of Orange II solutions using heterogeneous catalysts based on saponite clay. *Appl. Catal. B* **2007**, *71*, 44–56. [CrossRef]
29. Navalon, S.; Alvaro, M.; Garcia, H. Heterogeneous Fenton catalysts based on clays, silicas and zeolites. *Appl. Catal. B* **2010**, *99*, 1–26. [CrossRef]
30. Hassan, H.; Hameed, B.H. Iron-clay as effective heterogeneous Fenton catalyst for the decolorization of Reactive Blue 4. *Chem. Eng. J.* **2011**, *171*, 912–918. [CrossRef]
31. Kuznetsova, E.V.; Savinov, E.N.; Vostrikova, L.A.; Parmon, V.N. Heterogeneous catalysis in the Fenton-type system FeZSM-5/H$_2$O$_2$. *Appl. Catal. B* **2004**, *51*, 165–170. [CrossRef]
32. Velichkova, F.; Delmas, H.; Julcour, C.; Koumanova, B. Heterogeneous fenton and photo-fenton oxidation for paracetamol removal using iron containing ZSM-5 zeolite as catalyst. *AIChE J.* **2017**, *63*, 669–679. [CrossRef]
33. Ramirez, J.H.; Maldonado-Hodar, F.J.; Perez-Cadenas, A.F.; Moreno-Castilla, C.; Costa, C.A.; Madeira, L.M. Azo-dye Orange II degradation by heterogeneous Fenton-like reaction using carbon-Fe catalysts. *Appl. Catal. B* **2007**, *75*, 312–323. [CrossRef]
34. Duarte, F.; Maldonado-Hodar, F.J.; Perez-Cadenas, A.F.; Madeira, L.M. Fenton-like degradation of azo-dye Orange II catalyzed by transition metals on carbon aerogels. *Appl. Catal. B* **2009**, *85*, 139–147. [CrossRef]
35. Sun, L.; Yao, Y.; Wang, L.; Mao, Y.; Huang, Z.; Yao, D.; Lu, W.; Chen, W. Efficient removal of dyes using activated carbon fibers coupled with 8-hydroxyquinoline ferric as a reusable Fenton-like catalyst. *Chem. Eng. J.* **2014**, *240*, 413–419. [CrossRef]
36. Wang, L.; Yao, Y.; Zhang, Z.; Sun, L.; Lu, W.; Chen, W.; Chen, H. Activated carbon fibers as an excellent partner of Fenton catalyst for dyes decolorization by combination of adsorption and oxidation. *Chem. Eng. J.* **2014**, *251*, 348–354. [CrossRef]
37. Carrasco-Diaz, M.R.; Castillejos-Lopez, E.; Cerpa-Naranjo, A.; Rojas-Cervantes, M.L. On the textural and crystalline properties of Fe-carbon xerogels. Application as Fenton-like catalysts in the oxidation of paracetamol by H$_2$O$_2$. *Microporous Mesoporous Mater.* **2017**, *237*, 282–293. [CrossRef]
38. Eberhardt, M.K.; Ramirez, G.; Ayala, E. Does the reaction of copper(I) with hydrogen peroxide give hydroxyl radicals? A study of aromatic hydroxylation. *J. Org. Chem.* **1989**, *54*, 5922–5926. [CrossRef]

39. Lassmann, G.; Eriksson, L.A.; Himo, F.; Lendzian, F.; Lubitz, W. Electronic Structure of a Transient Histidine Radical in Liquid Aqueous Solution: EPR Continuous-Flow Studies and Density Functional Calculations. *J. Phys. Chem. A* **1999**, *103*, 1283–1290. [CrossRef]
40. Li, H.; Wan, J.; Ma, Y.; Wang, Y.; Chen, X.; Guan, Z. Degradation of refractory dibutyl phthalate by peroxymonosulfate activated with novel catalysts cobalt metal-organic frameworks: Mechanism, performance and stability. *J. Hazard. Mater.* **2016**, *318*, 154–163. [CrossRef] [PubMed]
41. Rivas, F.J.; Carbajo, M.; Beltran, F.J.; Acedo, B.; Gimeno, O. Perovskite catalytic ozonation of pyruvic acid in water. *Appl. Catal. B* **2006**, *62*, 93–103. [CrossRef]
42. Carbajo, M.; Rivas, F.J.; Beltran, F.J.; Alvarez, P.; Medina, F. Effects of different catalysts on the ozonation of pyruvic acid in water. *Ozone Sci. Eng.* **2006**, *28*, 229–235. [CrossRef]
43. Carbajo, M.; Beltran, F.J.; Medina, F.; Gimeno, O.; Rivas, F.J. Catalytic ozonation of phenolic compounds. *Appl. Catal. B* **2006**, *67*, 177–186. [CrossRef]
44. Carbajo, M.; Beltran, F.J.; Gimeno, O.; Acedo, B.; Rivas, F.J. Ozonation of phenolic wastewaters in the presence of a perovskite type catalyst. *Appl. Catal. B* **2007**, *74*, 203–210. [CrossRef]
45. Beltran, F.J.; Pocostales, P.; Alvarez, P.M.; Lopez-Pineiro, F. Catalysts to improve the abatement of sulfamethoxazole and the resulting organic carbon in water during ozonation. *Appl. Catal. B* **2009**, *92*, 262–270. [CrossRef]
46. Beltran, F.J.; Pocostales, P.; Alvarez, P.; Garcia-Araya, J.F.; Gimeno, O. Perovskite catalytic ozonation of some pharmaceutical compounds in water. *Ozone Sci. Eng.* **2010**, *32*, 230–237. [CrossRef]
47. Orge, C.A.; Orfao, J.J.M.; Pereira, M.F.R.; Barbero, B.P.; Cadus, L.E. Lanthanum-based perovskites as catalysts for the ozonation of selected organic compounds. *Appl. Catal. B* **2013**, *140-141*, 426–432. [CrossRef]
48. Afzal, S.; Quan, X.; Zhang, J. High surface area mesoporous nanocast LaMO$_3$ (M = Mn, Fe) perovskites for efficient catalytic ozonation and an insight into probable catalytic mechanism. *Appl. Catal. B* **2017**, *206*, 692–703. [CrossRef]
49. Zhang, Y.; Xia, Y.; Li, Q.; Qi, F.; Xu, B.; Chen, Z. Synchronously degradation benzotriazole and elimination bromate by perovskite oxides catalytic ozonation: Performance and reaction mechanism. *Sep. Purif. Technol.* **2018**, *197*, 261–270. [CrossRef]
50. Beltran, F.J.; Aguinaco, A.; Garcia-Araya, J.F.; Oropesa, A.L. Ozone and photocatalytic processes to remove the antibiotic sulfamethoxazole from water. *Water Res.* **2008**, *42*, 3799–3808. [CrossRef] [PubMed]
51. Rivas, F.J.; Carbajo, M.; Beltran, F.; Gimeno, O.; Frades, J. Comparison of different advanced oxidation processes (AOPs) in the presence of perovskites. *J. Hazard. Mater.* **2008**, *155*, 407–414. [CrossRef] [PubMed]
52. De Laat, J.; Gallard, H. Catalytic Decomposition of hydrogen peroxide by Fe(III) in homogeneous aqueous solution: Mechanism and kinetic modeling. *Environ. Sci. Technol.* **1999**, *33*, 2726–2732. [CrossRef]
53. Bokare, A.D.; Choi, W. Review of iron-free Fenton-like systems for activating H_2O_2 in advanced oxidation processes. *J. Hazard. Mater.* **2014**, *275*, 121–135. [CrossRef] [PubMed]
54. Luo, W.; Zhu, L.; Wang, N.; Tang, H.; Cao, M.; She, Y. Efficient removal of organic pollutants with magnetic nanoscaled BiFeO$_3$ as a reusable heterogeneous Fenton-like catalyst. *Environ. Sci. Technol.* **2010**, *44*, 1786–1791. [CrossRef] [PubMed]
55. Zhang, L.; Nie, Y.; Hu, C.; Qu, J. Enhanced Fenton degradation of Rhodamine B over nanoscaled Cu-doped LaTiO$_3$ perovskite. *Appl. Catal. B* **2012**, *125*, 418–424. [CrossRef]
56. Xiao, P.; Hong, J.; Wang, T.; Xu, X.; Yuan, Y.; Li, J.; Zhu, J. Oxidative degradation of organic dyes over supported perovskite oxide LaFeO$_3$/SBA-15 under ambient conditions. *Catal. Lett.* **2013**, *143*, 887–894. [CrossRef]
57. Xiao, P.; Li, H.; Wang, T.; Zhu, J. Efficient Fenton-like La-Cu-O/SBA-15 catalyst for the degradation of organic dyes under ambient conditions. *RSC Adv.* **2014**, *4*, 12601–12604. [CrossRef]
58. Li, H.; Zhu, J.; Xiao, P.; Zhan, Y.; Lv, K.; Wu, L.; Li, M. On the mechanism of oxidative degradation of rhodamine B over LaFeO$_3$ catalysts supported on silica materials: Role of support. *Microporous Mesoporous Mater.* **2016**, *221*, 159–166. [CrossRef]
59. Giuseppina Iervolinoa, V.V.; Sanninoa, D.; Rizzob, L.; Sarnoa, G.; Lyubov, P.C.; Isupovac, A. Influence of operating conditions in the photo-Fenton removal of tartrazine on structured catalysts. *Chem. Eng. Trans.* **2015**, *43*, 979–984.
60. Zhao, H.; Cao, J.; Lv, H.; Wang, Y.; Zhao, G. 3D nano-scale perovskite-based composite as Fenton-like system for efficient oxidative degradation of ketoprofen. *Catal. Commun.* **2013**, *41*, 87–90. [CrossRef]

61. Li, J.; Miao, J.; Duan, X.; Dai, J.; Liu, Q.; Wang, S.; Zhou, W.; Shao, Z. Fine-tuning surface properties of perovskites via nanocompositing with inert oxide toward ceveloping superior catalysts for advanced oxidation. *Adv. Funct. Mater.* **2018**, *28*, 1804654. [CrossRef]
62. Nie, Y.; Zhang, L.; Li, Y.-Y.; Hu, C. Enhanced Fenton-like degradation of refractory organic compounds by surface complex formation of $LaFeO_3$ and H_2O_2. *J. Hazard. Mater.* **2015**, *294*, 195–200. [CrossRef] [PubMed]
63. Rusevova, K.; Koeferstein, R.; Rosell, M.; Richnow, H.H.; Kopinke, F.-D.; Georgi, A. $LaFeO_3$ and $BiFeO_3$ perovskites as nanocatalysts for contaminant degradation in heterogeneous Fenton-like reactions. *Chem. Eng. J.* **2014**, *239*, 322–331. [CrossRef]
64. Taran, O.P.; Ayusheev, A.B.; Ogorodnikova, O.L.; Prosvirin, I.P.; Isupova, L.A.; Parmon, V.N. Perovskite-like catalysts $LaBO_3$ (B = Cu, Fe, Mn, Co, Ni) for wet peroxide oxidation of phenol. *Appl. Catal. B* **2016**, *180*, 86–93. [CrossRef]
65. Ben Hammouda, S.; Zhao, F.; Safaei, Z.; Babu, I.; Ramasamy, D.L.; Sillanpaa, M. Reactivity of Ceria-Perovskite composites CeO_2-$LaMO_3$ (M=Cu, Fe) in catalytic wet peroxidative oxidation of pollutant Bisphenol F: Characterization, kinetic and mechanism studies. *Appl. Catal. B* **2017**, *218*, 119–136. [CrossRef]
66. Wang, N.; Zhu, L.; Lei, M.; She, Y.; Cao, M.; Tang, H. Ligand-induced drastic enhancement of catalytic activity of nano-$BiFeO_3$ for oxidative degradation of bisphenol A. *ACS Catal.* **2011**, *1*, 1193–1202. [CrossRef]
67. Song, Z.; Wang, N.; Zhu, L.; Huang, A.; Zhao, X.; Tang, H. Efficient oxidative degradation of triclosan by using an enhanced Fenton-like process. *Chem. Eng. J.* **2012**, *198-199*, 379–387. [CrossRef]
68. Magalhaes, F.; Moura, F.C.C.; Ardisson, J.D.; Lago, R.M. $LaMn_{1-x}Fe_xO_3$ and $LaMn_{0.1-x}Fe_{0.90}Mo_xO_3$ perovskites: Synthesis, characterization and catalytic activity in H_2O_2 reactions. *Mater. Res.* **2008**, *11*, 307–312. [CrossRef]
69. Jauhar, S.; Dhiman, M.; Bansal, S.; Singhal, S. Mn^{3+} ion in perovskite lattice: A potential Fenton's reagent exhibiting remarkably enhanced degradation of cationic and anionic dyes. *J. Sol-Gel Sci. Technol.* **2015**, *75*, 124–133. [CrossRef]
70. Carrasco-Diaz, M.R.; Castillejos-Lopez, E.; Cerpa-Naranjo, A.; Rojas-Cervantes, M.L. Efficient removal of paracetamol using $LaCu_{1-x}M_xO_3$ (M = Mn, Ti) perovskites as heterogeneous Fenton-like catalysts. *Chem. Eng. J.* **2016**, *304*, 408–418. [CrossRef]
71. Sannino, D.; Vaiano, V.; Ciambelli, P.; Isupova, L.A. Mathematical modelling of the heterogeneous photo-Fenton oxidation of acetic acid on structured catalysts. *Chem. Eng. J.* **2013**, *224*, 53–58. [CrossRef]
72. Loghambal, S.; Agvinos Catherine, A.J. Mathematical analysis of the heterogeneous photo-Fenton oxidation of acetic acid on structured catalysts. *J. Math. Chem.* **2016**, *54*, 1146–1158. [CrossRef]
73. Singh, C.; Rakesh, M. Oxidation of phenol using $LaMnO_3$ perovskite, TiO_2, H_2O_2 and UV radiation. *Indian J. Chem. Technol.* **2010**, *17*, 451–454.
74. Soltani, T.; Lee, B.-K. Enhanced formation of sulfate radicals by metal-doped $BiFeO_3$ under visible light for improving photo-Fenton catalytic degradation of 2-chlorophenol. *Chem. Eng. J.* **2017**, *313*, 1258–1268. [CrossRef]
75. Ju, L.; Chen, Z.; Fang, L.; Dong, W.; Zheng, F.; Shen, M. Sol-gel synthesis and photo-Fenton-like catalytic activity of $EuFeO_3$ nanoparticles. *J. Am. Ceram. Soc.* **2011**, *94*, 3418–3424. [CrossRef]
76. Phan, T.T.N.; Nikoloski, A.N.; Bahri, P.A.; Li, D. Heterogeneous photo-Fenton degradation of organics using highly efficient Cu-doped $LaFeO_3$ under visible light. *J. Ind. Eng. Chem.* **2018**, *61*, 53–64. [CrossRef]
77. Sannino, D.; Vaiano, V.; Ciambelli, P.; Isupova, L.A. Structured catalysts for photo-Fenton oxidation of acetic acid. *Catal. Today* **2011**, *161*, 255–259. [CrossRef]
78. Sannino, D.; Vaiano, V.; Isupova, L.A.; Ciambelli, P. Heterogeneous photo-Fenton oxidation of organic pollutants on structured catalysts. *J. Adv. Oxid. Technol.* **2012**, *15*, 294–300. [CrossRef]
79. Vaiano, V.; Isupova, L.A.; Ciambelli, P.; Sannino, D. Photo-fenton oxidation of t-butyl methyl ether in presence of $LaFeO_3$ supported on monolithic structure. *J. Adv. Oxid. Technol.* **2014**, *17*, 187–192. [CrossRef]
80. Orak, C.; Atalay, S.; Ersoz, G. Photocatalytic and photo-Fenton-like degradation of methylparaben on monolith-supported perovskite-type catalysts. *Sep. Sci. Technol.* **2017**, *52*, 1310–1320. [CrossRef]
81. An, J.; Zhu, L.; Wang, N.; Song, Z.; Yang, Z.; Du, D.; Tang, H. Photo-Fenton like degradation of tetrabromobisphenol A with graphene $BiFeO_3$ composite as a catalyst. *Chem. Eng. J.* **2013**, *219*, 225–237. [CrossRef]

82. Kondrakov, A.O.; Ignatev, A.N.; Lunin, V.V.; Frimmel, F.H.; Braese, S.; Horn, H. Roles of water and dissolved oxygen in photocatalytic generation of free OH radicals in aqueous TiO_2 suspensions: An isotope labeling study. *Appl. Catal. B* **2016**, *182*, 424–430. [CrossRef]
83. Chiou, C.-S.; Chen, Y.-H.; Chang, C.-T.; Chang, C.-Y.; Shie, J.-L.; Li, Y.-S. Photochemical mineralization of di-n-butyl phthalate with H_2O_2/Fe^{3+}. *J. Hazard. Mater.* **2006**, *135*, 344–349. [CrossRef] [PubMed]
84. Li, X.; Liu, Y.; Wang, C.; Yin, K. Study of the stability of H_2O_2 in alkaline slurry. *Bandaoti Jishu* **2012**, *37*, 850–854.
85. Ovejero, G.; Sotelo, J.L.; Martinez, F.; Gordo, L. Novel heterogeneous catalysts in the wet peroxide oxidation of phenol. *Water Sci. Technol.* **2001**, *44*, 153–160. [CrossRef] [PubMed]
86. Sotelo, J.L.; Ovejero, G.; Martinez, F.; Melero, J.A.; Milieni, A. Catalytic wet peroxide oxidation of phenolic solutions over a $LaTi_{1-x}Cu_xO_3$ perovskite catalyst. *Appl. Catal. B* **2004**, *47*, 281–294. [CrossRef]
87. Faye, J.; Guelou, E.; Barrault, J.; Tatibouet, J.M.; Valange, S. $LaFeO_3$ Perovskite as new and performant catalyst for the wet peroxide oxidation of organic pollutants in ambient conditions. *Top. Catal.* **2009**, *52*, 1211–1219. [CrossRef]
88. Guerra-Rodríguez, S.; Rodríguez, E.; Singh, N.D.; Rodríguez-Chueca, J. Assessment of Sulfate Radical-Based Advanced Oxidation Processes for Water and Wastewater Treatment: A Review. *Water* **2018**, *10*, 1828. [CrossRef]
89. Ghanbari, F.; Moradi, M. Application of peroxymonosulfate and its activation methods for degradation of environmental organic pollutants: Review. *Chem. Eng. J.* **2017**, *310*, 41–62. [CrossRef]
90. Khan, A.; Liao, Z.; Liu, Y.; Jawad, A.; Ifthikar, J.; Chen, Z. Synergistic degradation of phenols using peroxymonosulfate activated by $CuO-Co_3O_4$@MnO_2 nanocatalyst. *J. Hazard. Mater.* **2017**, *329*, 262–271. [CrossRef] [PubMed]
91. Duan, X.; Ao, Z.; Sun, H.; Indrawirawan, S.; Wang, Y.; Kang, J.; Liang, F.; Zhu, Z.H.; Wang, S. Nitrogen-doped graphene for generation and evolution of reactive radicals by metal-free catalysis. *ACS Appl. Mater. Interfaces* **2015**, *7*, 4169–4178. [CrossRef] [PubMed]
92. Anipsitakis, G.P.; Dionysiou, D.D.; Gonzalez, M.A. Cobalt-mediated activation of peroxymonosulfate and sulfate radical attack on phenolic compounds. Implications of chloride ions. *Environ. Sci. Technol.* **2006**, *40*, 1000–1007. [CrossRef] [PubMed]
93. Sun, H.; Liang, H.; Zhou, G.; Wang, S. Supported cobalt catalysts by one-pot aqueous combustion synthesis for catalytic phenol degradation. *J. Colloid Interface Sci.* **2013**, *394*, 394–400. [CrossRef] [PubMed]
94. Ji, Y.; Dong, C.; Kong, D.; Lu, J. New insights into atrazine degradation by cobalt catalyzed peroxymonosulfate oxidation: Kinetics, reaction products and transformation mechanisms. *J. Hazard. Mater.* **2015**, *285*, 491–500. [CrossRef] [PubMed]
95. Ben Hammouda, S.; Zhao, F.; Safaei, Z.; Srivastava, V.; Lakshmi Ramasamy, D.; Iftekhar, S.; kalliola, S.; Sillanpää, M. Degradation and mineralization of phenol in aqueous medium by heterogeneous monopersulfate activation on nanostructured cobalt based-perovskite catalysts $ACoO_3$ (A = La, Ba, Sr and Ce): Characterization, kinetics and mechanism study. *Appl. Catal. B* **2017**, *215*, 60–73. [CrossRef]
96. Su, C.; Duan, X.; Miao, J.; Zhong, Y.; Zhou, W.; Wang, S.; Shao, Z. Mixed conducting perovskite materials as superior catalysts for fast aqueous-phase advanced oxidation: A mechanistic study. *ACS Catal.* **2017**, *7*, 388–397. [CrossRef]
97. Lin, K.-Y.A.; Chen, Y.-C.; Lin, T.-Y.; Yang, H. Lanthanum cobaltite perovskite supported on zirconia as an efficient heterogeneous catalyst for activating Oxone in water. *J. Colloid Interface Sci.* **2017**, *497*, 325–332. [CrossRef] [PubMed]
98. Miao, J.; Zhou, W.; Shao, Z.; Sunarso, J.; Su, C.; Wang, S. $SrCo_{1-x}Ti_xO_{3-\delta}$ perovskites as excellent catalysts for fast degradation of water contaminants in neutral and alkaline solutions. *Sci. Rep.* **2017**, *7*, 44215. [CrossRef] [PubMed]
99. Duan, X.; Su, C.; Miao, J.; Zhong, Y.; Shao, Z.; Wang, S. Sun, H. Insights into perovskite-catalyzed peroxymonosulfate activation: Maneuverable cobalt sites for promoted evolution of sulfate radicals. *Appl. Catal. B* **2018**, *220*, 626–634. [CrossRef]
100. Lu, S.; Wang, G.; Chen, S.; Yu, H.; Ye, F.; Quan, X. Heterogeneous activation of peroxymonosulfate by $LaCo_{1-x}Cu_xO_3$ perovskites for degradation of organic pollutants. *J. Hazard. Mater.* **2018**, *353*, 401–409. [CrossRef] [PubMed]

101. Miao, J.; Sunarso, J.; Duan, X.; Zhou, W.; Wang, S.; Shao, Z. Nanostructured Co-Mn containing perovskites for degradation of pollutants: Insight into the activity and stability. *J. Hazard. Mater.* **2018**, *349*, 177–185. [CrossRef] [PubMed]

102. Pang, X.; Guo, Y.; Zhang, Y.; Xu, B.; Qi, F. LaCoO$_3$ perovskite oxide activation of peroxymonosulfate for aqueous 2-phenyl-5-sulfobenzimidazole degradation: Effect of synthetic method and the reaction mechanism. *Chem. Eng. J.* **2016**, *304*, 897–907. [CrossRef]

103. RSolis, R.; Rivas, F.J.; Gimeno, O. Removal of aqueous metazachlor, tembotrione, tritosulfuron and ethofumesate by heterogeneous monopersulfate decomposition on lanthanum-cobalt perovskites. *Appl. Catal. B* **2017**, *200*, 83–92. [CrossRef]

104. Lin, K.-Y.A.; Chen, Y.-C.; Lin, Y.-F. LaMO$_3$ perovskites (M=Co, Cu, Fe and Ni) as heterogeneous catalysts for activating peroxymonosulfate in water. *Chem. Eng. Sci.* **2017**, *160*, 96–105. [CrossRef]

105. Rao, Y.; Zhang, Y.; Han, F.; Guo, H.; Huang, Y.; Li, R.; Qi, F.; Ma, J. Heterogeneous activation of peroxymonosulfate by LaFeO$_3$ for diclofenac degradation: DFT-assisted mechanistic study and degradation pathways. *Chem. Eng. J.* **2018**, *352*, 601–611. [CrossRef]

106. Chu, Y.; Tan, X.; Shen, Z.; Liu, P.; Han, N.; Kang, J.; Duan, X.; Wang, S.; Liu, L.; Liu, S. Efficient removal of organic and bacterial pollutants by Ag-La$_{0.8}$Ca$_{0.2}$Fe$_{0.94}$O$_{3-\delta}$ perovskite via catalytic peroxymonosulfate activation. *J. Hazard. Mater.* **2018**, *356*, 53–60. [CrossRef] [PubMed]

107. Chi, F.; Song, B.; Yang, B.; Lv, Y.; Ran, S.; Huo, Q. Activation of peroxymonosulfate by BiFeO$_3$ microspheres under visible light irradiation for decomposition of organic pollutants. *RSC Adv.* **2015**, *5*, 67412–67417. [CrossRef]

108. Solis, R.R.; Rivas, F.J.; Gimeno, O.; Perez-Bote, J.-L. Synergism between peroxymonosulfate and LaCoO$_3$-TiO$_2$ photocatalysis for oxidation of herbicides. Operational variables and catalyst characterization assessment. *J. Chem. Technol. Biotechnol.* **2017**, *92*, 2159–2170. [CrossRef]

109. Ben Hammouda, S.; Zhao, F.; Safaei, Z.; Ramasamy, D.L.; Doshi, B.; Sillanpaa, M. Sulfate radical-mediated degradation and mineralization of bisphenol F in neutral medium by the novel magnetic Sr$_2$CoFeO$_6$ double perovskite oxide catalyzed peroxymonosulfate: Influence of co-existing chemicals and UV irradiation. *Appl. Catal. B* **2018**, *233*, 99–111. [CrossRef]

110. Fendrich, A.M.; Quaranta, A.; Orlandi, M.; Bettonte, M.; Miotello, A. Solar Concentration for Wastewaters Remediation: A Review of Materials and Technologies. *Appl. Sci.* **2018**, *9*, 118. [CrossRef]

111. Li, S.; Jing, L.; Fu, W.; Yang, L.; Xin, B.; Fu, H. Photoinduced charge property of nanosized perovskite-type LaFeO$_3$ and its relationships with photocatalytic activity under visible irradiation. *Mater. Res. Bull.* **2007**, *42*, 203–212. [CrossRef]

112. Su, H.J.; Jing, L.Q.; Shi, K.Y.; Yao, C.H.; Fu, H.G. Synthesis of large surface area LaFeO$_3$ nanoparticles by SBA-16 template method as high active visible photocatalysts. *J. Nanopart. Res.* **2010**, *12*, 967–974. [CrossRef]

113. Hu, R.; Li, C.; Wang, X.; Sun, Y.; Jia, H.; Su, H.; Zhang, Y. Photocatalytic activities of LaFeO$_3$ and La$_2$FeTiO$_6$ in p-chlorophenol degradation under visible light. *Catal. Commun.* **2012**, *29*, 35–39. [CrossRef]

114. Li, L.; Wang, X.; Zhang, Y. Enhanced visible light-responsive photocatalytic activity of LnFeO$_3$ (Ln = La, Sm) nanoparticles by synergistic catalysis. *Mater. Res. Bull.* **2014**, *50*, 18–22. [CrossRef]

115. Hou, L.; Sun, G.; Liu, K.; Li, Y.; Gao, F. Preparation, characterization and investigation of catalytic activity of Li-doped LaFeO$_3$ nanoparticles. *J. Sol-Gel Sci. Technol.* **2006**, *40*, 9–14. [CrossRef]

116. Shen, H.; Xue, T.; Wang, Y.; Cao, G.; Lu, Y.; Fang, G. Photocatalytic property of perovskite LaFeO$_3$ synthesized by sol-gel process and vacuum microwave calcination. *Mater. Res. Bull.* **2016**, *84*, 15–24. [CrossRef]

117. Tang, P.; Tong, Y.; Chen, H.; Cao, F.; Pan, G. Microwave-assisted synthesis of nanoparticulate perovskite LaFeO$_3$ as a high active visible-light photocatalyst. *Curr. Appl. Phys.* **2013**, *13*, 340–343. [CrossRef]

118. Li, F.-t.; Liu, Y.; Liu, R.-h.; Sun, Z.-m.; Zhao, D.-s.; Kou, C.-g. Preparation of Ca-doped LaFeO$_3$ nanopowders in a reverse microemulsion and their visible light photocatalytic activity. *Mater. Lett.* **2010**, *64*, 223–225. [CrossRef]

119. Thirumalairajan, S.; Girija, K.; Ganesh, I.; Mangalaraj, D.; Viswanathan, C.; Balamurugan, A.; Ponpandian, N. Controlled synthesis of perovskite LaFeO3 microsphere composed of nanoparticles via self-assembly process and associated photocatalytic activity. *Chem. Eng. J.* **2012**, *209*, 420–428. [CrossRef]

120. Ding, J.; Lue, X.; Shu, H.; Xie, J.; Zhang, H. Microwave-assisted synthesis of perovskite ReFeO$_3$ (Re: La, Sm, Eu, Gd) photocatalyst. *Mater. Sci. Eng. B* **2010**, *171*, 31–34. [CrossRef]

121. Liang, Q.; Jin, J.; Liu, C.; Xu, S.; Li, Z. Constructing a novel p-n heterojunction photocatalyst LaFeO$_3$/g-C$_3$N$_4$ with enhanced visible-light-driven photocatalytic activity. *J. Alloy. Compd.* **2017**, *709*, 542–548. [CrossRef]
122. Ren, X.; Yang, H.; Sai, G.; Zhou, J.; Yang, T.; Zhang, X.; Cheng, Z.; Sun, S. Controlled growth of LaFeO$_3$ nanoparticles on reduced graphene oxide for highly efficient photocatalysis. *Nanoscale* **2016**, *8*, 752–756. [CrossRef] [PubMed]
123. Fu, S.; Niu, H.; Tao, Z.; Song, J.; Mao, C.; Zhang, S.; Chen, C.; Wang, D. Low temperature synthesis and photocatalytic property of perovskite-type LaCoO$_3$ hollow spheres. *J. Alloys Compd.* **2013**, *576*, 5–12. [CrossRef]
124. Song, S.; Xu, L.; He, Z.; Chen, J.; Xiao, X.; Yan, B. Mechanism of the Photocatalytic Degradation of C.I. Reactive Black 5 at pH 12.0 Using SrTiO$_3$/CeO$_2$ as the Catalyst. *Environ. Sci. Technol.* **2007**, *41*, 5846–5853. [CrossRef] [PubMed]
125. Song, S.; Xu, L.J.; He, Z.Q.; Ying, H.P.; Chen, J.M.; Xiao, X.Z.; Yan, B. Photocatalytic degradation of CI Direct Red 23 in aqueous solutions under UV irradiation using SrTiO$_3$/CeO$_2$ composite as the catalyst. *J. Hazard. Mater.* **2008**, *152*, 1301–1308. [CrossRef] [PubMed]
126. Huerta-Flores, A.M.; Sanchez-Martinez, D.; del Rocio Hernandez-Romero, M.; Zarazua-Morin, M.E.; Torres-Martinez, L.M. Visible-light-driven BaBiO$_3$ perovskite photocatalysts: Effect of physicochemical properties on the photoactivity towards water splitting and the removal of rhodamine B from aqueous systems. *J. Photochem. Photobiol. A* **2019**, *368*, 70–77. [CrossRef]
127. Zheng, H.X.; Zhang, T.T.; Zhu, Y.M.; Liang, B.; Jiang, W. KBiO$_3$ as an Effective Visible-Light-Driven Photocatalyst: Degradation Mechanism for Different Organic Pollutants. *Chemphotochem* **2018**, *2*, 442–449. [CrossRef]
128. Sun, S.; Wang, W.; Zhang, L.; Shang, M. Visible Light-Induced Photocatalytic Oxidation of Phenol and Aqueous Ammonia in Flowerlike Bi$_2$Fe$_4$O$_9$ Suspensions. *J. Phys. Chem. C* **2009**, *113*, 12826–12831. [CrossRef]
129. Gao, F.; Chen, X.; Yin, K.; Dong, S.; Ren, Z.; Yuan, F.; Yu, T.; Zou, Z.G.; Liu, J.-M. Visible-light photocatalytic properties of weak magnetic BiFeO$_3$ nanoparticles. *Adv. Mater.* **2007**, *19*, 2889–2892. [CrossRef]
130. Huang, H.; Lu, B.; Liu, Y.; Wang, X.; Hu, J. Synthesis of LaMnO$_3$-Diamond Composites and Their Photocatalytic Activity in the Degradation of Weak Acid Red C-3GN. *Nano* **2018**, *13*, 1850121. [CrossRef]
131. Chasse, M.; Hallett-Tapley, G.L. Gold nanoparticle-functionalized niobium oxide perovskites as photocatalysts for visible light-induced aromatic alcohol oxidations. *Can. J. Chem.* **2018**, *96*, 664–671. [CrossRef]
132. Ohno, T.; Tsubota, T.; Nakamura, Y.; Sayama, K. Preparation of S, C cation-codoped SrTiO$_3$ and its photocatalytic activity under visible light. *Appl. Catal. A-Gen.* **2005**, *288*, 74–79. [CrossRef]
133. Dong, B.; Li, Z.C.; Li, Z.Y.; Xu, X.R.; Song, M.X.; Zheng, W.; Wang, C.; Al-Deyab, S.S.; El-Newehy, M. Highly Efficient LaCoO$_3$ Nanofibers Catalysts for Photocatalytic Degradation of Rhodamine, B. *J. Am. Ceram. Soc.* **2010**, *93*, 3587–3590. [CrossRef]
134. Qin, C.; Li, Z.; Chen, G.; Zhao, Y.; Lin, T. Fabrication and visible-light photocatalytic behavior of perovskite praseodymium ferrite porous nanotubes. *J. Power Sources* **2015**, *285*, 178–184. [CrossRef]
135. Xiang, Q.; Lang, D.; Shen, T.; Liu, F. Graphene-modified nanosized Ag$_3$PO$_4$ photocatalysts for enhanced visible-light photocatalytic activity and stability. *Appl. Catal. B* **2015**, *162*, 196–203. [CrossRef]
136. Guo, J.; Dai, Y.-z.; Chen, X.-j.; Zhou, L.-l.; Liu, T.-h. Synthesis and characterization of Ag$_3$PO$_4$/LaCoO$_3$ nanocomposite with superior mineralization potential for bisphenol A degradation under visible light. *J. Alloys Compd.* **2017**, *696*, 226–233. [CrossRef]
137. Gao, K.; Li, S. Multi-modal TiO$_2$–LaFeO$_3$ composite films with high photocatalytic activity and hydrophilicity. *Appl. Surf. Sci.* **2012**, *258*, 6460–6464. [CrossRef]
138. Dhinesh Kumar, R.; Thangappan, R.; Jayavel, R. Synthesis and characterization of LaFeO$_3$/TiO$_2$ nanocomposites for visible light photocatalytic activity. *J. Phys. Chem. Solids* **2017**, *101*, 25–33. [CrossRef]
139. Schunemann, S.; van Gastel, M.; Tuysuz, H.A. CsPbBr$_3$/TiO$_2$ Composite for Visible-Light-Driven Photocatalytic Benzyl Alcohol Oxidation. *Chemsuschem* **2018**, *11*, 2057–2061. [CrossRef] [PubMed]
140. Li, G.; Zhang, Y.; Wu, L.; Wu, F.; Wang, R.; Zhang, D.; Zhu, J.; Li, H. An efficient round-the-clock La$_2$NiO$_4$ catalyst for breaking down phenolic pollutants. *RSC Adv.* **2012**, *2*, 4822–4828. [CrossRef]
141. Sun, M.; Jiang, Y.; Li, F.; Xia, M.; Xue, B.; Liu, D. Dye degradation activity and stability of perovskite-type LaCoO$_{3-x}$ (x = 0∼0.075). *Mater. Trans.* **2010**, *51*, 2208–2214. [CrossRef]

142. Tummino, M.L.; Laurenti, E.; Deganello, F.; Bianco Prevot, A.; Magnacca, G. Revisiting the catalytic activity of a doped SrFeO$_3$ for water pollutants removal: Effect of light and temperature. *Appl. Catal. B* **2017**, *207*, 174–181. [CrossRef]
143. Wu, J.-M.; Wen, W. Catalyzed degradation of azo dyes under ambient conditions. *Environ. Sci. Technol.* **2010**, *44*, 9123–9127. [CrossRef] [PubMed]
144. Zhong, W.; Jiang, T.; Dang, Y.; He, J.; Chen, S.-Y.; Kuo, C.-H.; Kriz, D.; Meng, Y.; Meguerdichian, A.G.; Suib, S.T. Mechanism studies on Methyl Orange dye degradation by perovskite-type LaNiO$_{3-\delta}$ under dark ambient conditions. *Appl. Catal. A* **2018**, *549*, 302–3099. [CrossRef]
145. Leiw, M.Y.; Guai, G.H.; Wang, X.; Tse, M.S.; Ng, C.M.; Tan, O.K. Dark ambient degradation of Bisphenol A and Acid Orange 8 as organic pollutants by perovskite SrFeO$_{3-\delta}$ metal oxide. *J. Hazard. Mater.* **2013**, *260*, 1–8. [CrossRef] [PubMed]
146. Chen, H.; Motuzas, J.; Martens, W.; Diniz da Costa, J.C. Degradation of azo dye Orange II under dark ambient conditions by calcium strontium copper perovskite. *Appl. Catal. B* **2018**, *221*, 691–700. [CrossRef]

© 2019 by the authors. Licensee MDPI, Basel, Switzerland. This article is an open access article distributed under the terms and conditions of the Creative Commons Attribution (CC BY) license (http://creativecommons.org/licenses/by/4.0/).

Article

Fe-Doping in Double Perovskite PrBaCo$_{2(1-x)}$Fe$_{2x}$O$_{6-\delta}$: Insights into Structural and Electronic Effects to Enhance Oxygen Evolution Catalyst Stability

Bae-Jung Kim [1,*], **Emiliana Fabbri** [1,*], **Ivano E. Castelli** [2], **Mario Borlaf** [3], **Thomas Graule** [3], **Maarten Nachtegaal** [1] **and Thomas J. Schmidt** [1,4]

1. Paul Scherrer Institut, Forschungstrasse 111, 5232 Villigen PSI, Switzerland;
   maarten.nachtegaal@psi.ch (M.N.); thomasjustus.schmidt@psi.ch (T.J.S.)
2. Department of Energy Conversion and Storage, Fysikvej 309, Technical University of Denmark,
   DK-2800 Kgs. Lyngby, Denmark; ivca@dtu.dk
3. Laboratory for High Performance Ceramics, Empa, Swiss Federal Laboratories for Materials Testing and
   Research, 8600 Dübendorf, Switzerland; mario.borlaf@empa.ch (M.B.); thomas.graule@empa.ch (T.G.)
4. Laboratory of Physical Chemistry, ETH Zürich, CH-8093 Zürich, Switzerland
* Correspondence: joseph.kim@psi.ch (B.-J.K.); emiliana.fabbri@psi.ch (E.F.);
  Tel.: +41-56-310-4580 (B.K.); +41-56-310-2795 (E.F.)

Received: 19 February 2019; Accepted: 12 March 2019; Published: 14 March 2019

Abstract: Perovskite oxides have been gaining attention for its capability to be designed as an ideal electrocatalyst for oxygen evolution reaction (OER). Among promising candidates, the layered double perovskite—PrBaCo$_2$O$_{6-\delta}$ (PBC)—has been identified as the most active perovskite electrocatalyst for OER in alkaline media. For a single transition metal oxide catalyst, the addition of Fe enhances its electrocatalytic performance towards OER. To understand the role of Fe, herein, Fe is incorporated in PBC in different ratios, which yielded PrBaCo$_{2(1-x)}$Fe$_{2x}$Co$_{6-\delta}$ (x = 0, 0.2 and 0.5). Fe-doped PBCF's demonstrate enhanced OER activities and stabilities. Operando X-ray absorption spectroscopy (XAS) revealed that Co is more stable in a lower oxidation state upon Fe incorporation by establishing charge stability. Hence, the degradation of Co is inhibited such that the perovskite structure is prolonged under the OER conditions, which allows it to serve as a platform for the oxy(hydroxide) layer formation. Overall, our findings underline synergetic effects of incorporating Fe into Co-based layered double perovskite in achieving a higher activity and stability during oxygen evolution reaction.

Keywords: Fe-substitution; operando X-ray absorption spectroscopy; oxygen evolution reaction; double perovskite catalysts; oxy(hydroxide)

1. Introduction

Today, modern society is evolving to become more energy dependent. As the awareness of environmental impact from current energy systems is elevating, more efforts recently have been devoted towards mainstreaming renewable energy sources. However, the implementation of renewable energy technologies is challenging, as it requires an efficient energy storage system to mediate the intermittent generation and consumption of energy. The electrochemical splitting of water (i.e., water electrolysis) offers an effective method to produce large amounts of hydrogen (H$_2$), which can be stored and used as an energy vector [1]. Therefore, an efficient oxygen evolution reaction (OER) is essential since it is the key reaction in water electrolysis. During the past decade, the members of the perovskite oxide family (ABO$_3$) have been gaining vast attention for their promising activities as OER electrocatalysts under alkaline conditions, and thereby relieving the need of expensive precious metals such as iridium [2–7]. Generally, perovskite oxides are composed of rare-earth (e.g., lanthanides) or earth alkaline metals (e.g., Ba) in the A-site and 3D transition metals in the B-site (e.g., Ni, Co and/or

Fe). The perovskites intrinsic properties can be tailored through partial cation substitution [2,8,9]. This substitution can transpire in both A- and B-sites of the perovskite (i.e., $A'_{1-n}A''_{n}B'_{1-m}B''_{m}O_3$; $0 \leq n, m \leq 1$) either in ordered or random arrangement [10]. Likewise, cation ordering plays an important role in engineering the intrinsic properties of a perovskite such as electronic structure, ionic conductivity, and magnetic properties, all of which may change its electrocatalytic behavior [3,10,11]. The recently proposed OER mechanism emphasizes that the formation of oxy(hydroxide) layer at metal oxide surface is essential along the path of lattice oxygen evolution reaction (LOER) [12,13]. In this context, developing perovskite oxides as OER catalyst is advantageous, owing to its ability to exhibit high oxygen vacancy concentration upon cation substitution and ordering it so as to activate the LOER [14,15]. In contrast to the conventional OER mechanism [16,17], in the case where the lattice oxygen is directly involved (i.e., LOER), the high surface OH^- coverage from the alkaline media is no longer necessary as loosely bonded lattice oxygen atoms act as the reaction intermediates itself; as a result, the overpotential is lowered [12,18]. In this regard, the use of layered double perovskite oxides ($A'_{2(1-n)}A''_{2n}B'_{2(1-m)}B''_{2m}O_6$; $0 \leq n, m \leq 1$) is beneficial as they tend to localize the oxygen vacancies into layers through A-site ordering and promote high oxygen mobility [14]. Among the layered double perovskite family, $PrBaCo_2O_{6-\delta}$ (PBC) has been appraised for its high OER activity [19–24]. Nevertheless, past studies point out the instability of PBC under OER conditions [2,13,23,25–27], raising queries regarding its degradation mechanism. In our recent study [2], we highlighted that the degradation is a kinetic process and every catalyst varies in how it reaches the end of its service life depending on the inherent properties.

In a different perspective, a single randomly ordered perovskite oxide, $Ba_{0.5}Sr_{0.5}Co_{0.8}Fe_{0.2}O_{3-\delta}$ (BSCF), has been identified as another highly active OER catalyst [5,22,27,28]. More recently, we have reported that highly oxygen deficient BSCF prepared via flame spray synthesis would lead to the participation of lattice oxygen atoms (i.e., LOER) coupled with the OER process. Based on operando X-ray absorption spectroscopy (XAS) results [2,24,27] and density-functional theory (DFT) based calculations [2,24], BSCF is capable of facilitating the formation of a self-constructed oxy(hydroxide) surface layer during OER, owing to its thermodynamic nature of meta-stability under the OER condition. Intriguingly, in the means to understand each individual chemical component of BSCF, recent findings highlight the vital role of Fe in its B-site so as to pertain to the thermodynamic meta-stability, and provide charge stability [24]. Likewise, many studies have reported constructive effect of incorporating Fe into 3D transition metal oxide catalysts for OER [29–38].

Therefore, in this study, we incorporate Fe into the B-site of PBC in different ratios to yield $PrBaCo_{2(1-x)}Fe_{2x}Co_{6-\delta}$ (x = 0.2 and 0.5; denoted as PBCF82 and PBCF55, respectively) for the purpose of tailoring the electrocatalytic performance with respect to OER activity and stability. Nanoparticles of all the materials under study are attained via flame spray synthesis [39]. Operando XAS is used to gain insights into changes in local electronic and geometric structures of the layered double perovskites upon Fe-doping. Combined with the thermodynamic nature inferred from DFT calculations, the roles of Fe in the layered double perovskite as OER catalyst are highlighted. Based on our findings, we underline the synergetic effect that Fe conveys and elucidate the enhanced OER performance of layered double perovskite catalyst upon Fe incorporation.

2. Results and Discussion

2.1. Physcial Characterization

All perovskite catalysts—$PrBaCo_{2(1-x)}Fe_{2x}Co_{6-\delta}$ (x = 0, 0.2, and 0.5; denoted as PBC, PBCF82, and PBCF55, respectively)—are prepared via flame spray synthesis from which nano-scaled particles are obtained (Figure 1). Metal precursors are dissolved in combustible solution and injected into the flame at high temperature (possible up to ≈3000 °C) and the resulting precipitates are collected. Previously, we have established the benefits of this particular synthesis of perovskites for the electrocatalytic performance [2,27]. The physical and structural traits of such nanoparticles are observed using transmission electron microscopy (TEM) (Figure 1). Figure 1a–c show the TEM images of the prepared nanoparticles of PBC, PBCF82, and PBCF55, respectively. All of the prepared nanoparticles are in sizes that range from 5–30 nm. Each inset of Figure 1a–c shows high-resolution (HR) TEM images of PBC, PBCF82, and PBCF55, respectively, and each reveals clear fringes, indicating the formation of crystalline structures.

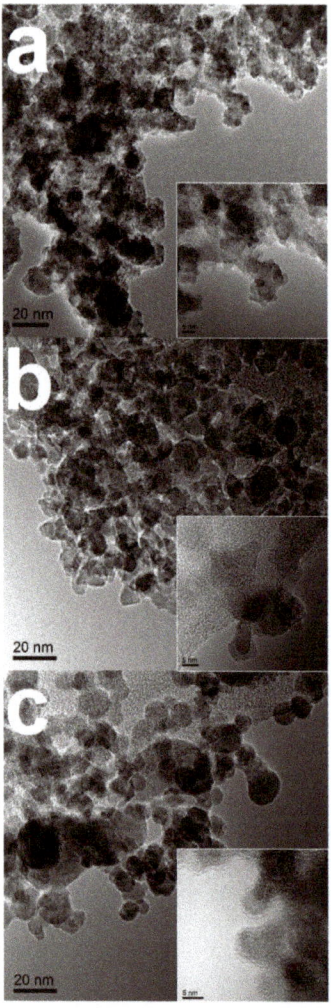

Figure 1. Transmission electron microscopy (TEM) images of (**a**) PBC; (**b**) PBCF82; and (**c**) PBCF55. Each inset shows high-resolution TEM images revealing fringes.

Figure 2 shows the comparison of X-ray diffraction (XRD) patterns of the prepared layered double perovskite catalysts. The comparison of the overall XRD patterns (Figure 2a) of PBC, PBCF82, and PBCF55 reveal peaks that are well indexed to those characteristics of $PrBaCo_2O_{6-\delta}$ (P4/mmm) (JCPDS 00-053-0131) with a minor amount of side oxide phases. The broad XRD peaks render convolution of nearby peaks, and confirms the presence of nanoparticles as observed in TEM images. Apart from being in the same crystalline structure, the XRD peaks of PBCF82 and PBCF55 appear at a lower 2-theta value (Figure 2b) than PBC, which highlights that the perovskite structure exhibits larger lattice parameters upon the incorporation of Fe. The peaks of PBCF55—the one with the most Fe composition—appears at the lowest 2-theta value, suggesting that it exhibits the largest lattice parameters among them. Considering the nature of ordered perovskites, and which cations are oriented in layers, the partial substitution of Co with Fe may lead to B-site cation octahedral tilting, which is difficult to separate from cation ordering [8]. Overall, the comparison of XRD patterns confirms that the layered double perovskite structure is withheld upon Fe incorporation while lattices may expand.

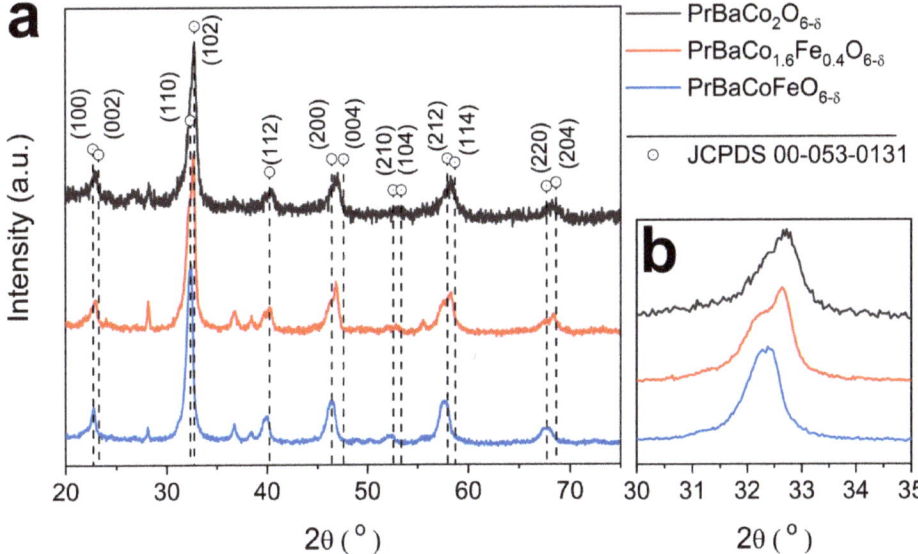

Figure 2. (**a**) comparison of X-ray diffraction (XRD) patterns of PBC (black), PBCF82 (red), and PBCF55 (blue). The XRD patterns are well indexed to those of $PrBaCo_2O_{6-\delta}$ from literature (JCPDS 00-053-0131); (**b**) magnified view of the main XRD peak.

2.2. Electrochemical Study

Systematic electrochemical characterizations are conducted in order to assess the functional role of Fe in activity and stability of the layered double perovskite catalysts during the OER process. Figure 3a shows constructed Tafel plots from measured steady-state currents from series of chronoamperometry studies of the prepared catalysts. In the presence of Fe, PBCF82 and PBCF55 revealed different Tafel slopes as compared to the non-doped PBC. PBCF82 and PBCF55 both showed similar Tafel slopes (~50 mV dec^{-1}) that are lower than that of the non-doped PBC (72 mV dec^{-1}) (summarized in Table 1; CVs and steady-state current recorded during the series of chronoamperometry measurement is shown in Figure S1). Our recent study investigated the functional role of Fe in single randomly ordered Co-based perovskites, $Ba_{0.5}Sr_{0.5}CoO_{3-\delta}$ and $La_{0.2}Sr_{0.8}CoO_{3-\delta}$, where remarkable enhancements in their OER activities were reported when doped with ~5 wt.% of Fe [24]. However, in this previous study, comparable Tafel slopes were observed for both non-doped and Fe-doped perovskites [24]. In contrast, here, the observed lower Tafel slope indicates that the catalyst would follow a different OER mechanism upon Fe incorporation, which may attribute to certain degree of changes of intrinsic physicochemical properties. PBCF82 and PBCF55 reveal higher current densities at 1.55 V_{RHE} (17.1 and 19.7 A g^{-1}, respectively) than compared to the non-doped PBC (13.8 A g^{-1}). Likewise, lower Tafel slopes have been observed for Co based catalysts with increasing Fe composition in other studies [29,31,40]. All of the prepared layer perovskites revealed similar Brunauer–Emmett–Teller (BET) surface areas (Table S1), which shows that their differences in electrochemical activities are independent of their differences in surface area.

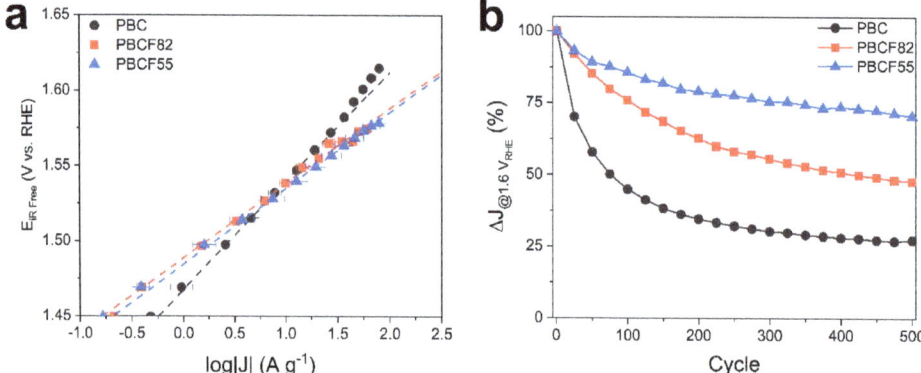

Figure 3. Electrochemical study comparing (**a**) Tafel plot of oxygen evolution reaction activities and (**b**) change in current densities with respect to the initial current density at every 25 cycles over 500 cycles between 1.0 and 1.6 V_{RHE} of PBC (black), PBCF82 (red), and PBCF55 (blue).

As previous studies emphasize the importance of understanding the stabilities of electrode materials [2,41–43], the potential stability test was conducted during which steady-state currents are recorded as potential is stepped from 1.0 V_{RHE} to 1.6 V_{RHE} holding for 10 seconds at each potential for 500 cycles (Figure 3b). Here, the Fe-doping also showed a beneficial effect in functional stability where PBCF55 revealed the least amount of current density loss over the period of cycles (lost 32% of its initial current density), while PBC lost about 74%. PBCF82, which comprises less than half of Fe that is in PBCF55, demonstrated capability to retain about half of its initial current density at the end of the cycle (see Table 1). In this comparison, the most outstanding activity and stability are demonstrated by PBCF55 containing one-to-one ratio between Co:Fe in the B-site.

Table 1. Summary of electrochemical study results on oxygen evolution reaction activity and stability: apparent mass specific exchange current density (j_0), Tafel slope (b), activity expressed as mass specific current density at 1.54 V, and stability expressed as percent of initial mass-specific current density after the 500 cycles of stepping between 1.0 V_{RHE} and 1.6 V_{RHE}.

Catalysts	j_0 (A g^{-1})	OER Tafel slope (mV dec^{-1})	j @ 1.55 V_{RHE} (A g^{-1})	Stability △J (%)
PBC	1.47	72	13.8	−73%
PBCF82	1.49	50	17.1	−52%
PBCF55	1.48	50	19.7	−32%

The functional stability of perovskite catalyst is described based on the ability to sustain its initial electrocatalytic activity. In relation to the structural integrity, the catalyst should demonstrate capabilities to serve as a suitable substrate for the prospect of surface oxy(hydroxide) layer formation [24]. In this regard, Pourbaix diagrams are constructed based on density-functional theory (DFT) calculations to assess the thermodynamically stable phases of PBC and PBCF in aqueous solution (Figure S2a,b, respectively). The Pourbaix diagrams of PBC and PBCF suggest the dissolution of A- and B-site cations. In fact, the cation dissolutions of perovskites are pointed out to be inevitable as governed by thermodynamics under the necessary OER conditions [41]. Nevertheless, it is essential to highlight that the degradation mechanism is also driven by kinetics. Hence, each perovskite catalyst would reach the end of its service life through different pathways based on the defect chemistry under the OER conditions [2,12,43]. In the light of above findings, the incorporation of Fe seems to impede the rate of degradation, where the one with the highest Fe composition (i.e., PBCF55) showed the most enhanced potential stability. Nonetheless, it should be noted that a higher composition of Fe (i.e., Fe-rich; x > 0.5) may compromise the role of Co as the active center and intrinsically undergo a different OER mechanism than Co-rich PBCF (i.e., x < 0.5) which would dim the assessment of the functional role of Fe-doping [31,40].

2.3. Operando X-ray Absorption Spectroscopy Study

The results of electrochemical studies revealed increasing OER activity and stability with increasing Fe composition in the domain of $0 \leq x \leq 0.5$. As mentioned, $PrBaCo_{2(1-x)}Fe_{2x}Co_{6-\delta}$ with different Co to Fe composition ratios would follow different reaction pathways as they undergo different degradation processes under the OER condition. Therefore, operando X-ray absorption spectroscopy study is conducted in order to monitor the changes in local electronic and geometric structures of the prepared layered double perovskites during the OER process, through which the functional effects of Fe were highlighted.

In Figure 4a, Co K-edge energy positions of normalized X-ray absorption near edge structure (XANES) spectra of as-prepared double perovskites reveal that PBC and PBCFs have Co oxidation state between +2 and +3, for which the Co K-edge position of PBCF55 is positioned at the lowest energy level. Through comparison, $PrBaCo_{2(1-x)}Fe_{2x}Co_{6-\delta}$ show reduced Co oxidation states with increasing Fe composition (i.e., x) in the following descending order: PBC > PBCF82 > PBCF55, each with ~0.3 eV difference in their Co K-edge energy position. With this information, the concentration of oxygen vacancy can be relatively estimated as listed in ascending order: PBC < PBCF82 < PBCF55 (refer to Figure S3). Note that all edge energy positions were determined at the half of the edge step. Figure 4b shows the comparison of the Fourier-transformed (FT) k^3-weighted Co K-edge extended X-ray absorption fine structure (EXAFS) spectra of as-prepared PBC, PBCF82, and PBCF55. The peaks of FT-EXAFS spectra signify the presence of neighboring atomic shells at specific radial distances from the absorbing atom (i.e., Co). In Figure 4b, the first two major peaks at ~1.9 Å and ~2.8 Å are ascribable to the backscattering contributions from the nearest Co–O and Co–Co/Fe ligands, respectively. Here, it is noteworthy that the second peak is ascribable to Co–Co/Fe coordination shell of the edge-sharing polyhedra typically found in highly oxygen deficient perovskite oxides prepared by flame spray synthesis (see Figure S2) [2,44,45]. Inconveniently, this Co–Co/Fe ligand distance of edge-sharing polyhedra of $PrBaCo_{2(1-x)}Fe_{2x}Co_{6-\delta}$ is in close vicinity to that of Co/Fe-oxy(hydroxide) (refer to Supplementary Information S5 for detailed explanation). The next appearing peaks located at ~3.5 Å corresponds to Pr/Ba neighbors. All of the FT-EXAFS profiles of PBC, PBCF82, and PBCF55 show similar peak locations to one another, suggesting that a similar local structure is maintained upon Fe incorporation. In Figure 4b, the first peak amplitude is observed to be decreased in the presence of Fe; listed in the descending order of amplitude: PBC > PBCF82 > PBCF55. Given that the first peak is ascribable to Co–O coordination shell, Co of the layered double perovskite seems to be bound to less oxygen atoms at that radial distance in the presence of Fe. This is further verified by the best fit of the Co–O peak of FT-EXAFS spectra of as-prepared catalysts, which shows the decrease in Co–O coordination number with a higher amount of Fe-doping (refer to Figure S10). Together with the comparison of XANES spectra, these findings lead to assert that more oxygen vacant sites are created with a higher amount of Fe-doping, and therefore reduces Co oxidation state in $PrBaCo_{2(1-x)}Fe_{2x}Co_{6-\delta}$. Moreover, the decrease of scattering intensities at farther radial distances observed in the FT-EXAFS spectra of both PBCF82 and PBCF55 is rationalized by octahedral distortions induced by doping of Fe (+3) into the B-site replacing Co (~+2.7) cations, which then weakens the backscattering from the neighboring atoms [35,46–53].

Furthermore, Figure 4c,d display comparisons of normalized XANES and FT-EXAFS spectra, respectively, of PBCF82 and PBCF55 recorded at the Fe K-edge. In Figure 4c, the edge energy positions of Fe K-edge XANES spectra of PBCF82 and PBCF55 indicates that their Fe oxidation states are similar (between +3 and +4). More precisely, the Fe K-edge of PBCF55 is positioned at about ~0.1 eV lower energy than PBCF82, but this insignificant difference would make the comparison trivial. Figure 4d shows Fe K-edge FT-EXAFS spectra of both PBCF82 and PBCF55 with similar scattering patterns as those of Co K-edge FT-EXAFS spectra. This confirms that Fe is indeed well integrated into the B-site of perovskite structure.

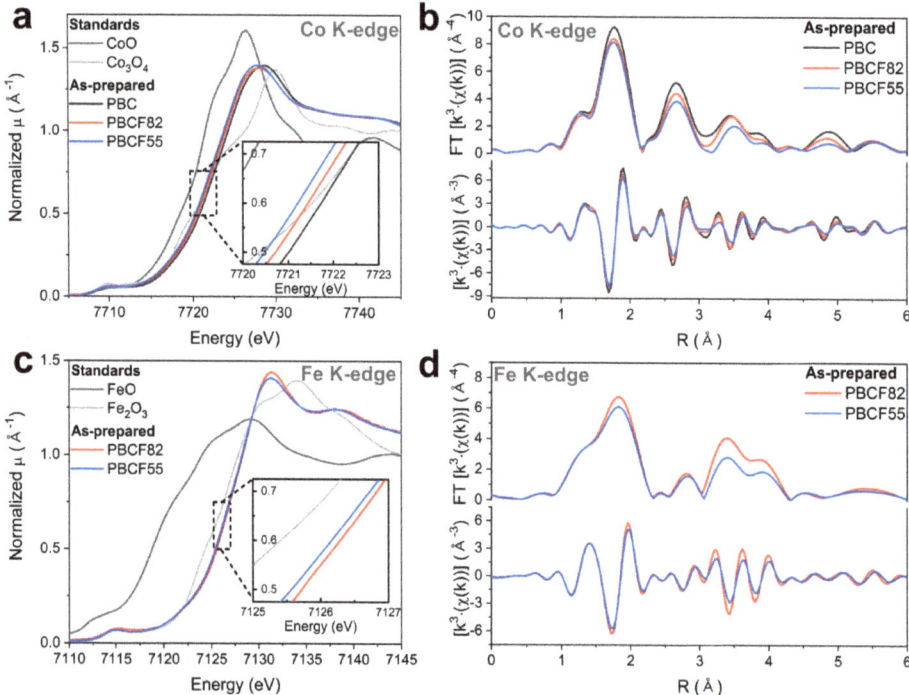

Figure 4. Comparison of as-prepared PBC, PBCF82, and PBCF55 Co K-edge (**a**) X-ray absorption near edge spectra (XANES) spectra; (**b**) Fourtier-transformed extended X-ray absorption fine structure (FT-EXAFS) spectra; and Fe K-edge; (**c**) XANES spectra; and (**d**) FT-EXAFS spectra.

When a neutral oxygen vacancy is created during synthesis, the left behind electrons would be distributed to either to Co and/or Fe, which would lower their oxidation states. However, the electrons are not evenly distributed among them but rather more accepted by Fe [40,48]. Based on above evidence, the addition of Fe seems to aid Co to be stable in a lower oxidation state by providing charge stabilization through balancing oxygen non-stoichiometry, and therefore promotes formation of oxygen vacancies. In the presence of higher oxygen vacancy concentration, the lattice oxygen within the perovskite structure is more inclined to participate in the water oxidation reaction (i.e., LOER) and develops the OER active oxy(hydroxide) layer at the surface [3,27].

Figure 5a shows a shift of the Co K-edge XANES spectra to higher energy positions of PBC, PBCF82, and PBCF55 during the operando flow cell test, among which PBCF55 reveals the greatest extent of energy shift of ~0.7 eV at the highest potential, while PBC and PBCF82 display ~0.3 eV of edge shifts. Our previous studies showed that the positive Co K-edge shift during the OER process is attributed to the construction of the OER active oxy(hydroxide) layer, since the Co oxidation state of Co-oxy(hydroxide) layer is higher than those inherent in the perovskite oxide [2,27]. This also agrees with other DFT based studies [46,47,51]. Therefore, revisiting Figure 3a, the high OER activity of

PBCF55 can be explained by the large extent of increase in the Co oxidation state reading from the shift of the Co K-edge position. At this point, we emphasize that the rate of dissolution (i.e., degradation) of a perovskite oxide catalyst during the OER process is controlled by its kinetics; thereby, each catalyst would reach its end-of-service life at a different rate. In this context, Figure 5b shows Co K-edge shifts at each increasing applied potential with respect to the energy positions recorded at 1.2 V_{RHE} during the anodic polarization in flow cell test. In Figure 5b, Fe-doped PBCF82 and PBCF55 reveal rapid positive Co K-edge shifts when polarized into the oxygen evolution regime (> 1.4 V_{RHE}). The edge shift translates proportionally to the increase of Co oxidation state. This sudden increase of Co oxidation state when polarized above the OER onset signifies formation of the OER active oxy(hydroxide) layer, which is potential-induced. Meanwhile, in the absence of Fe, PBC shows a consistent increase even at potentials below the oxygen evolution regime, indicating that the oxidation of its Co species is triggered by chemical dissolution [2].

Particularly, even though the extent of Co K-edge shift of PBCF82 is smaller than that of PBCF55, PBCF82 demonstrated a similar OER activity as PBCF55. Nevertheless, PBCF82 showed a rapid yet subtle increase of the Co oxidation state when polarized above the OER onset as similar to PBCF55 (Figure 5b). This may suggest the development of Co oxy(hydroxide) layer occurs but still less than in the case of PBCF55. Based on these observations, PBCF82 and PBCF55 both show different Co oxidation behaviors than compared to PBC, suggesting that the catalyst would undergo a modified degradation pathway during the OER process upon Fe incorporation. Referring to the Tafel plot (Figure 3a), the difference in their Tafel slopes, where lower Tafel slopes are observed upon Fe-doping, may suggest that PBCF's would undergo different mechanism during OER than the non-doped PBC. Also referring to the stability test (Figure 3b), the inhibited degradation mechanism upon Fe incorporation is further supported by the improved current stabilities of PBCF82 and PBCF55. Although thermodynamics anticipate the dissolution of $PrBaCo_{2(1-x)}Fe_{2x}Co_{6-\delta}$, it is intriguing to observe an enhanced ability to retain the initial current density upon Fe-doping, where PBCF55 demonstrated the highest retention of current density over the course of 500 cycles. The above findings point out that Fe plays an important role in improving the stability owing to the retardation of degradation mechanism, where the potential-induced increase of Co oxidation state is more attributable to the development of Co-oxy(hydroxide) species. However, this does not mean that deterioration of the perovskite structure is completely avoided as the decreasing trend of current densities are observed during the stability test. In brief, while the rate of chemical dissolution is lagged upon the addition of Fe, the layered double perovskite can be sustained as a substrate for the development of OER active oxy(hydroxide) species.

Despite the clear indications as to which the construction of Co-oxy(hydroxide) is displayed by positive edge shifts in their Co K-edge XANES spectra, this development—along with other concurrent local structural changes—are not clearly manifested in the comparison of FT-EXAFS spectra collected at potentials below and above the OER onset (1.2 and 1.54 V_{RHE}, respectively) (Figure S6a–c). Here, it is important to recapitulate the coinciding Co–Co radial distances between the edge-sharing polyhedra of $PrBaCo_{2(1-x)}Fe_{2x}Co_{6-\delta}$ and that of the Co-oxy(hydroxide) layer, both of which are at the proximity of ~2.8 Å from their primary Co atoms. In this respect, the FT-EXAFS profiles with these concurring signals would confound the precise interpretation of local structural changes. Considering these challenges, only the first peak is fitted in order to verify the changes in the Co–O coordination during the OER process (summarized in Table S3 and Figure S10). The best fits show increased Co–O coordination numbers for all catalysts at the highest anodic potential (1.54 V_{RHE}) and therefore further consolidates the increase of Co oxidation states during the OER process.

While the increase of Co oxidation states is evident, Fe K-edge XANES spectra collected during the operando measurement (Figure 5c) reveal insignificant edge shifts (~0.1 eV) throughout the course of anodic polarization. The Fe in both PBCF82 and PBCF55 are in a higher oxidation state (between +3 and +4) than Co (< +2.7), which allows Co to be in a more reduced state and thereby being more flexible in accommodating charge transfers [48]; this justifies the insignificant shift in the Fe K-edge.

In this context, the formation of the active oxy(hydroxide) layer would therefore be accommodated more by Co polyhedra, and thereby less changes would be manifested in their Fe K-edge FT-EXAFS profiles (Figure S6).

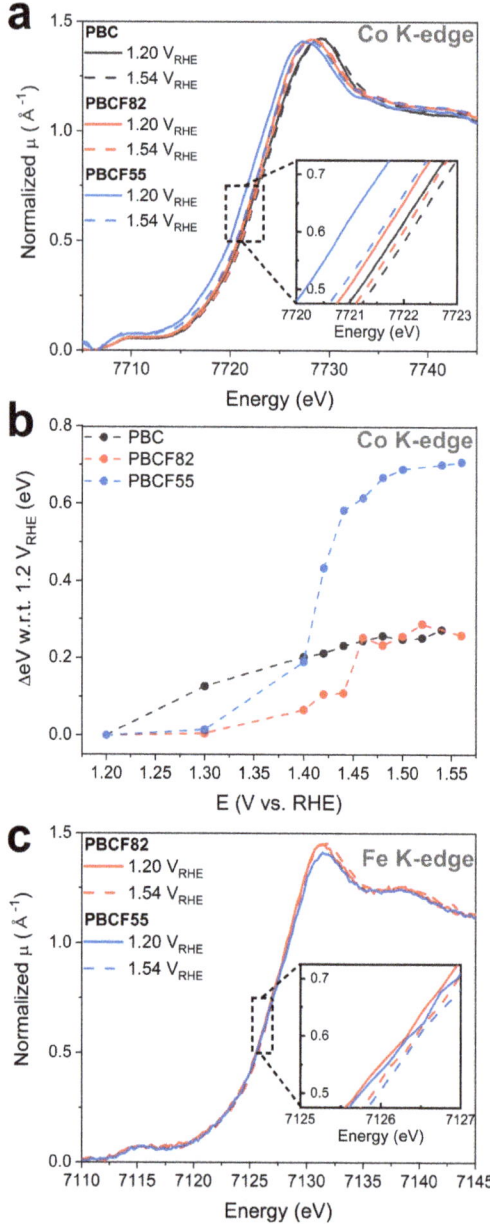

Figure 5. (**a**) comparison of Co K-edge XANES spectra of PBC, PBCF82, and PBCF55 recorded at 1.2 and 1.54 V_{RHE}; and (**b**) Co K-edge energy shift measured with respect to the edge position at 1.2 V_{RHE} at each potential during operando flow cell tests; (**c**) comparison of Fe K-edge XANES spectra of PBCF82 and PBCF55 recorded at 1.2 and 1.54 V_{RHE}.

In light of our findings, the Fe incorporation into the layered double perovskite catalyst leads to stabilizing Co in a lower oxidation state by providing a better charge distribution and promoting the formation of oxygen vacancies. Consequently, the degradation mechanism is inhibited such that the oxidation of Co is more induced by the increase of potential while its chemical dissolution is decelerated yet still unavoidable. The potential-induced Co oxidation behavior upon Fe-doping is indicative of the development of the OER active Co-oxy(hydroxide) surface layer. Despite these enhancements, no significant electronic and structural changes were detected with respect to Fe. Overall, the addition of Fe seems to enable the layered double perovskite catalyst to improve its structural integrity (similar as previously observed for $Ni_xFe_{1-x}O_2$) [37] as a more suitable substrate for the construction of oxy(hydroxide) layer leading to enhanced OER activity and stability.

3. Materials and Methods

3.1. Material Synthesis

The flame spray synthesis setup is described by Heel et al [54]. For the preparation of the $PrBaCo_2O_{5+x}$ (PBC) and $PrBaCo_{2-y}Fe_yO_{5+x}$ (y = 0.4 and 1.0) precursor solutions, stoichiometric amounts of praseodymium oxide (Pr_6O_{11}, 99.9%, Auer Remy, Hamburg, Germany), barium carbonate ($BaCO_3$, ≥99%, Sigma-Aldrich, Darmstadt, Germany), cobalt nitrate hexahydrate ($Co(NO_3)_2 \cdot 6H_2O$, 99.9%, Sigma-Aldrich, Darmstadt, Germany) and iron nitrate nonahydrate ($Fe(NO_3)_3 \cdot 9H_2O$, ≥98%, Sigma-Aldrich, Darmstadt, Germany) were dissolved in a mixture of solvents composed by N,N-Dimethylformamide (DMF, ≥99.8%, Roth, Frankfurt, Germany), acetic acid (HAc, ≥99.0%, Sigma-Aldrich, Darmstadt, Germany), nitric acid (HNO_3, 70%, Sigma-Aldrich, Darmstadt, Germany) and water in 45:25:5:25 volume ratio, respectively. Firstly, the Pr_6O_{11} was dissolved in the mixture of water and nitric acid at 80 °C; when a clear green solution was obtained, the BaCO3 was added and then, when no more CO_2 bubbles were observed, the Co and Fe (only for PBCF solutions) metal precursors. When all dissolved, the HAc and the DMF were added, obtaining a final total metal concentration of 0.1 M. The precursor solutions were pumped into the flame by using a three piston pump (C-601, Büchi, Flawil, Switzerland) with a flow controller (C-610, Büchi, Flawil, Switzerland), using a constant flow free of pulsations of 20 mL min^{-1}. Pure oxygen (99.95%, Carbagas, Bern, Switzerland) was used as dispersion gas with a flow rate of 35 L min^{-1}. The combustion gas was formed by acetylene (99.6%, Carbagas, Bern, Switzerland), with a flow rate of 13 L min^{-1}, and pure oxygen (17 L min^{-1}). Finally, the powders were collected in a baghouse filter on four ashless paper filters (Whatman, Buckinghamshire, UK).

3.2. Material Characterizations

Phase characterization of prepared materials was performed using powder X-ray diffraction (XRD, Bruker (Billerica, MA, USA) D8 system in Bragg–Brentano geometry, Cu K-α radiation (λ = 0.15418 nm)). Specific surface area was calculated by Brunauer–Emmett–Teller (BET) analysis of N_2 adsorption/desorption isotherms (AUROSORB-1, Quantachrome, Boynton Beach, FL, USA). Transmission electron microscopy (TEM) and energy dispersive X-ray spectroscopy (EDX) (TECNAI F30 operated at 300 kV) were used to study the surface morphology and composition of the prepared materials.

For ex situ and operando X-ray absorption spectroscopy (XAS) measurements, catalyst powders were dispersed in a mixture of isopropanol and Milli-Q water in the equal ratio sonicated for 30 min. The ink was then spray coated on Kapton film. XAS spectra at the Co K-edge and Fe K-edge were recorded at the SuperXAS beamline of the Swiss Light Source (PSI, Villigen, Switzerland). The incident photon beam provided by a 2.9 T superbend magnet source was collimated by a Si-coated mirror at 2.5 mRad (which also served to reject higher harmonics) and subsequently monochromatized by a Si (111) channel-cut monochromator. A Rh-coated toroidal mirror at 2.5 mRad was used to focus the X-ray beam to a spot size of 1 mm by 0.2 mm maximal on the sample position. The SuperXAS beamline [55] allowed for the rapid collection of 120 spectra during a measurement time of 60 sec (QEXAFS mode), which were then averaged. The spectra of samples were collected in transmission mode using N_2 filled ionization chambers, where a Co foil (of Fe foil) was placed between the second and third ionization chamber as a reference to calibrate and align collected spectra. Extended X-ray absorption fine structure (EXAFS) spectra were analyzed using the Demeter software package (0.9.26, Bruce Ravel, Washington D.C., USA) [56], which included background subtraction, energy calibration (based on the simultaneously measured Co reference foil) and edge step normalization. The resulting spectra were converted to the photoelectron wave vector k (in units $Å^{-1}$) by assigning the photoelectron energy origin, E_0, corresponding to k = 0, to the first inflection point of the absorption edge. The resulting $\chi(k)$ functions were weighted with k^3 to compensate for the dampening of the EXAFS amplitude. These $\chi(k)$ functions were Fourier transformed over 3–12 $Å^{-1}$ and plotted. The scattering paths used for the EXAFS fittings of all catalysts were generated from the structure of CoOOH [57] for the first Co–O coordination shell using the FEFF6.2 library. Refer to Supplementary Information S6 for detailed description of FT-EXAFS fittings.

3.3. Electrochemical Characterization

The electrochemical activities of the prepared catalysts were evaluated in a standard three-electrode electrochemical cell with the thin-film rotating disk electrode (RDE) [58]. The setup for OER and cyclic voltammetry (CV) consists of a potentiostat (Biologic VMP-300, Cary, NC, USA) and a rotation speed controlled motor (Pine Instrument Co., AFMSRCE, Grove City, PA, USA). All the electrochemical measurements were performed at standard room temperature using a reversible hydrogen electrode (RHE) as the reference electrode in 0.1 M KOH. A gold mesh was used as the counter electrode. A Teflon cell was used to contain the electrolyte with the electrodes immersed under potential control. A porous thin film electrode was prepared by drop-casting a catalyst ink on a polished glassy carbon electrode (5 mm OD/0.196 cm^2) [58]. The catalyst ink was prepared from a catalyst suspension made from sonicating (Bandelin, RM 16 UH, 300 Weff, 40 kHz, Berlin, Germany) 10 mg of catalyst in a solution mixture of 4 mL isopropyl alcohol (Sigma-Aldrich, 99.999%, Darmstadt, Germany) and 1 mL of Milli-Q water (ELGA, PURELAB Chorus, High Wycombe, UK), and 20 µL of Na^+-exchanged Nafion [4]. The 0.1 M KOH electrolyte was prepared by dissolving KOH pellets (Sigma-Aldrich, 99.99%, Darmstadt, Germany) in Milli-Q water.

First, 25 reverse potentiostatic sweeps of CV were conducted in synthetic air-saturated electrolyte from 1.0 to 1.7 V_{RHE} at a scan rate of 10 mV s^{-1}. Then, chronoamperometric measurements were carried out holding each potential step for 30 seconds to obtain steady-state currents at each potential from 1.2 to 1.7 V_{RHE} while rotating the working electrode at 900 rpm. The potential stability of catalyst materials was conducted using the same setup, by stepping potential between 1.0 and 1.6 V_{RHE} 500 times holding 10 seconds at each potential to collect currents at steady-state. Five cycles of CV and electrochemical impedance spectroscopy (EIS) were carried out after every 100 cycles. All measured currents were normalized by the mass of catalyst loading, and potentials were corrected for the ohmic drop obtained from EIS.

3.4. Operando Flow Cell Study

The homemade operando XAS electrochemical flow cell used in this study was extensively described previously [59]. As explained in Section 3.2, electrode materials were spray coated at the center of Kapton films. Black pearl-2000 (Cabot Corp., Boston, MA, USA) was used as the counter electrode material. A silver chloride electrode (Ag^+/AgCl) (Hugo Sachs Elektronik, March, Germany) was used as the reference electrode. During electrochemical testing, the electrolyte was drawn into the cell and collected in a syringe at the flow rate of 0.1 mL min^{-1}. The chronoamperometry measurement was carried out holding for 120 sec at each anodic potential step from 1.2 V_{RHE} to 1.54 V_{RHE}, and again during the reverse sequence back to 1.2 V_{RHE}. At each potential step, both transmission XAS spectra at

3.5. Density-Functional Theory—Pourbaix Diagrams

Density-functional theory (DFT) calculations are used to help understand the experimental data. In the first set of calculations, the stability of the perovskites in an aqueous environment was investigated by means of Pourbaix diagrams. This method has been previously used to investigate the stability of Ru-based perovskites in water as a descriptor for the stability in water to identify novel light harvesting materials [60]. Pourbaix diagrams show the phase diagram of solid and dissolved species as a function of pH and applied potential (V_{SHE}). DFT provides the total energies for the solid bulk for the perovskites and the other solid competing phases as described in the Materials Project database [61]. Experiments provide the dissolution energies for the dissolved species [62,63]. This method is implemented in the Atomic Simulation Environment (ASE) package [64], and more details about the methodology can be found in the literature [65,66]. All the bulk structures have been fully relaxed using the Quantum ESPRESSO code [67], PBEsol as exchange-correlation functional [68] and the pseudopotentials from the Standard Solid State Pseudopotential library (SSSP accuracy) [69]. We used Hubbard U+ correction of 4 eV and applied to Co, Fe, and Pr elements.

4. Conclusions

In here, a systematic study is conducted to assess the role of Fe in the highly OER active layered double perovskite catalyst, $PrBaCo_{2(1-x)}Fe_{2x}Co_{6-\delta}$, by comparing different compositions of Fe: PBC ($x = 0$), PBCF82 ($x = 0.2$), and PBCF55 ($x = 0.5$). These layered double perovskite catalysts are prepared via flame spray synthesis, where Fe is indeed incorporated into the B-site as verified by XRD and FT-EXAFS profiles. In comparison to PBC, Fe-doped PBCF82 and PBCF55 revealed enhanced OER activities and improved current stabilities as to better retain the initial current density. In the basis of our findings, such enhanced electrocatalytic performance is attributed to the addition of Fe, which provides charge stability as to compensate for the oxygen non-stoichiometry and allows Co to be in a lower oxidation state. This leads to alteration in the oxidation behavior of the layered double perovskite catalyst upon anodic polarization so that its Co oxidation is predominantly provoked by the increase of applied potential. In addition, the potential-induced Co oxidation upon Fe-doping is attributed to the formation of OER active oxy(hydroxide) layer at the surface. This also implies that the perovskite structure is prolonged under the OER conditions so that it serves as a substrate longer for the construction of oxy(hydroxide) layer at the surface. It should be noted that, even in the presence of Fe, the catalyst is not exempted from the inevitable cation dissolution under the OER conditions as governed by its thermodynamic nature. All of these effects lead to highlighting the constructive role of Fe in the layered double perovskite as an OER catalyst, which is evident from lower Tafel slopes and enhanced current density stabilities of PBCF82 and PBCF55. Particularly, PBCF55—with the highest Fe composition—demonstrated the greatest enhancement in both activity and stability as OER catalyst.

In summary, these results lead to concluding that a layered double perovskite catalyst is intrinsically modified upon Fe-doping so that it becomes more resilient to the cation dissolution, and thereby better supports the development of an oxy(hydroxide) surface layer. Such changes contribute to enhancements in the activity towards oxygen evolution reaction. As conventionally seen from many transition metal oxide catalysts, such electrocatalytic performance enhancements are attributed to the synergy between Fe and Co. In this regard, it should be emphasized that the incorporation of Fe would modify the degradation mechanism of the host transition metal. Thus, it should be generalized that the electrocatalytic performance of a doped-metal oxide would rely on the interaction between the host transition metal and the dopant metal. These findings present a systematic study method to engineer and design an ideal perovskite oxide as OER catalyst.

Supplementary Materials: The following are available online at http://www.mdpi.com/2073-4344/9/3/263/s1, Figure S1: The initial cyclic voltammetry scanned at 10 mV sec^{-1} from 1.0 to 1.7 V$_{RHE}$, and the series of chronoamperometry measurements recording steady-state current at each step of potential: (a–b) PBC; (c–d) PBCF82; and (e–f) PBCF55; Figure S2: Density-functional theory (DFT) calculated Pourbaix diagrams of (a) PrBaCo$_2$O$_{6-\delta}$ and (b) PrBaCo$_{2(1-x)}$Fe$_{2x}$O$_{6-\delta}$ (i.e., PBCF82 and PBCF55). Red lines mark the ranges of the working potential (on standard hydrogen electrode (SHE) potential scale) for the oxygen evolution reaction (OER) in at pH 13 used during the operando XAS study; Figure S3: Comparison of XANES spectra of PBC prepared via sol-gel (SG) method (black) and PBC (red), PBCF82 (blue), and PBCF55 (green) prepared via flame spray synthesis; Figure S4: Illustration of Co–Co distances in (a) Co octahedra, surrounded by six oxygen atoms in an octahedral structure sharing a single oxygen corner in a typical stoichiometric perovskite (ABO$_3$/A'A"B'B"O$_6$); and (b) B-site cation surrounded by less oxygen than stoichiometry (ABO$_{3-\delta}$/A'A"B'B"O$_{6-\delta}$); therefore, the network of polyhedra would stabilize by pivoting to share two oxygen atoms; Figure S5: Comparison of Fourier transformed (FT) k^3-weighted EXAFS profiles at Co K-edge of γ-Co-O(OH) [3] with as-prepared layered double perovskite catalysts: PBC (black), PBCF82 (red), and PBCF55 (blue); Figure S6: Comparison of FT Co K-edge EXAFS spectra collected at 1.20 and 1.54 V$_{RHE}$ during the operando XAS measurements of (a) PBC, (b) PBCF82, and (c) PBCF55. Simultaneously collected Fe K-edge EXAFS spectra of (d) PBCF82 and (e) PBCF55; Figure S7: Fourier transformed k$_3$-weighted Co K-edge EXAFS spectra of PBC (a) as-prepared, (b) at 1.2 V$_{RHE}$ anodic, and (c) 1.54 V$_{RHE}$ anodic. Black line is the FT-EXAFS spectrum, red line is the fitted spectrum, and blue is the window of the fitting; Figure S8: Fourier transformed k$_3$-weighted Co K-edge EXAFS spectra of PBCF82 (a) as-prepared, (b) at 1.2 V$_{RHE}$ anodic, and (c) 1.54 V$_{RHE}$ anodic. Black line is the FT-EXAFS spectrum, red line is the fitted spectrum, and blue is the window of the fitting; Figure S9: Fourier transformed k$_3$-weighted Co K-edge EXAFS spectra of PBCF55 (a) as-prepared, (b) at 1.2 V$_{RHE}$ anodic, and (c) 1.54 V$_{RHE}$ anodic. Black line is the FT-EXAFS spectrum, red line is the fitted spectrum, and blue is the window of the fitting; Figure S10: Comparison of coordination number (N$_{Co-O}$) of the first peak of FT-EXAFS spectra of PBC (black), PBCF82 (red), and PBCF55 (blue). Filled markers and empty markers represent NCo–O of the as-prepared catalysts and at 1.54 V$_{RHE}$ during the anodic polarization, respectively; Table S1: Summary of Brunauer–Emmet–Teller (BET) surface areas of the prepared layered double perovskite catalysts; Table S2: Lattice parameters of PBC, PBCF82, and PBCF55 calculated from Rietveld refinement of their X-ray diffractions. With the estimated lattice parameter, the Co–Co distance of edge-sharing polyhedron is calculated; Table S3: Summary of best fit parameters of the FT k$_3$-weighted Co K-edge EXAFS spectra of as-prepared and at 1.20 and 1.54 V$_{RHE}$ during the anodic polarization of (a) PBC, (b) PBCF82, and (c) PBCF55.

Author Contributions: Conception and conceptualization, B.-J.K., E.F., and T.J.S.; synthesis, M.B. and T.G.; characterizations, electrochemical measurements, and investigation, B.-J.K.; DFT computations, I.E.,C.; B.-J.K., E.F., I.E.,C., M.B., T.G., M.N. and T.J.S. discussed the results and commented on the manuscript.

Funding: This research was funded by Swiss National Science Foundation, Innosuisse and the Swiss Competence Center for Energy Research (SCCER) Heat and Paul Scherrer Institute.

Acknowledgments: The authors gratefully acknowledge the Swiss National Science Foundation through its Ambizione Program and the National Centre of Competence in Research (NCCR) Marvel, the Swiss Competence Center for Energy Research (SCCER) Heat and Electricity Storage through Innosuisse, Switzerland, and the Paul Scherrer Institute for financial contributions to this work, respectively. The authors thank the Swiss Light Source for providing beamtime at the SuperXAS beamline.

Conflicts of Interest: The authors declare no conflict of interest.

References

1. Babic, U.; Suermann, M.; Buehi, F.N.; Gubler, L.; Schmidt, T.J. Review-identifying critical gaps for polymer electrolyte water electrolysis development. *J. Electrochem. Soc.* **2017**, *164*, F387–F399. [CrossRef]
2. Kim, B.-J.; Cheng, X.; Abbott, D.F.; Fabbri, E.; Bozza, F.; Graule, T.; Castelli, I.E.; Wiles, L.; Danilovic, N.; Ayers, K.E.; et al. Highly active nanoperovskite catalysts for oxygen evolution reaction: Insights into activity and stability of ba0.5sr0.5co0.8fe0.2o2+δ and prbaco2o5+δ. *Adv. Funct. Mater.* **2018**, *28*, 1804355. [CrossRef]
3. Cheng, X.; Fabbri, E.; Yamashita, Y.; Castelli, I.E.; Kim, B.; Uchida, M.; Haumont, R.; Puente-Orench, I.; Schmidt, T.J. Oxygen evolution reaction on perovskites: A multieffect descriptor study combining experimental and theoretical methods. *ACS Catal.* **2018**, *8*, 9567–9578. [CrossRef]
4. Suntivich, J.; Gasteiger, H.A.; Yabuuchi, N.; Shao-Horn, Y. Electrocatalytic measurement methodology of oxide catalysts using a thin-film rotating disk electrode. *J. Electrochem. Soc.* **2010**, *157*, B1263–B1268. [CrossRef]
5. Grimaud, A.; May, K.J.; Carlton, C.E.; Lee, Y.L.; Risch, M.; Hong, W.T.; Zhou, J.G.; Shao-Horn, Y. Double perovskites as a family of highly active catalysts for oxygen evolution in alkaline solution. *Nat. Commun.* **2013**, *4*, 2439. [CrossRef] [PubMed]
6. Huang, X.B.; Zhao, G.X.; Wang, G.; Irvine, J.T.S. Synthesis and applications of nanoporous perovskite metal oxides. *Chem. Sci.* **2018**, *9*, 3623–3637. [CrossRef] [PubMed]

7. Hwang, J.; Rao, R.R.; Giordano, L.; Katayama, Y.; Yu, Y.; Shao-Horn, Y. Perovskites in catalysis and electrocatalysis. *Science* **2017**, *358*, 751–756. [CrossRef] [PubMed]
8. Vasala, S.; Karppinen, M. A2b'b"o6 perovskites: A review. *Prog. Solid State Chem.* **2015**, *43*, 1–36. [CrossRef]
9. Fabbri, E.; Habereder, A.; Waltar, K.; Kotz, R.; Schmidt, T.J. Developments and perspectives of oxide-based catalysts for the oxygen evolution reaction. *Catal. Sci. Technol.* **2014**, *4*, 3800–3821. [CrossRef]
10. King, G.; Woodward, P.M. Cation ordering in perovskites. *J. Mater. Chem.* **2010**, *20*, 5785–5796. [CrossRef]
11. Davies, P.K.; Wu, H.; Borisevich, A.Y.; Molodetsky, I.E.; Farber, L. Crystal chemistry of complex perovskites: New cation-ordered dielectric oxides. *Annu. Rev. Mater. Res.* **2008**, *38*, 369–401. [CrossRef]
12. Fabbri, E.; Schmidt, T.J. Oxygen evolution reaction—the enigma in water electrolysis. *ACS Catal.* **2018**, *8*, 9765–9774. [CrossRef]
13. Grimaud, A.; Diaz-Morales, O.; Han, B.H.; Hong, W.T.; Lee, Y.L.; Giordano, L.; Stoerzinger, K.A.; Koper, M.T.M.; Shao-Horn, Y. Activating lattice oxygen redox reactions in metal oxides to catalyse oxygen evolution. *Nat. Chem.* **2017**, *9*, 457–465. [CrossRef] [PubMed]
14. Taskin, A.A.; Lavrov, A.N.; Ando, Y. Achieving fast oxygen diffusion in perovskites by cation ordering. *Appl. Phys. Lett.* **2005**, *86*, 091910. [CrossRef]
15. Kim, G.; Wang, S.; Jacobson, A.J.; Yuan, Z.; Donner, W.; Chen, C.L.; Reimus, L.; Brodersen, P.; Mims, C.A. Oxygen exchange kinetics of epitaxial prbaco(2)o(5+delta) thin films. *Appl. Phys. Lett.* **2006**, *88*. [CrossRef]
16. Man, I.C.; Su, H.Y.; Calle-Vallejo, F.; Hansen, H.A.; Martinez, J.I.; Inoglu, N.G.; Kitchin, J.; Jaramillo, T.F.; Norskov, J.K.; Rossmeisl, J. Universality in oxygen evolution electrocatalysis on oxide surfaces. *ChemCatChem* **2011**, *3*, 1159–1165. [CrossRef]
17. Rossmeisl, J.; Qu, Z.W.; Zhu, H.; Kroes, G.J.; Norskov, J.K. Electrolysis of water on oxide surfaces. *J. Electroanal. Chem.* **2007**, *607*, 83–89. [CrossRef]
18. Damjanovic, A.; Jovanovic, B. Anodic oxide-films as barriers to charge-transfer in o2 evolution at pt in acid solutions. *J. Electrochem. Soc.* **1976**, *123*, 374–381. [CrossRef]
19. Han, B.; Shao-Horn, Y. (invited) in-situ study of the activated lattice oxygen redox reactions in metal oxides during oxygen evolution catalysis. *Meeting Abstracts* **2018**, *MA2018-01*, 1935.
20. Yoo, J.S.; Rong, X.; Liu, Y.; Kolpak, A.M. Role of lattice oxygen participation in understanding trends in the oxygen evolution reaction on perovskites. *ACS Catal.* **2018**, *8*, 4628. [CrossRef]
21. Hong, W.T.; Stoerzinger, K.A.; Lee, Y.L.; Giordano, L.; Grimaud, A.; Johnson, A.M.; Hwang, J.; Crumlin, E.J.; Yang, W.L.; Shao-Horn, Y. Charge-transfer-energy-dependent oxygen evolution reaction mechanisms for perovskite oxides. *Energ. Environ. Sci.* **2017**, *10*, 2190–2200. [CrossRef]
22. Suntivich, J.; May, K.J.; Gasteiger, H.A.; Goodenough, J.B.; Shao-Horn, Y. A perovskite oxide optimized for oxygen evolution catalysis from molecular orbital principles. *Science* **2011**, *334*, 1383–1385. [CrossRef] [PubMed]
23. Han, B.H.; Risch, M.; Lee, Y.L.; Ling, C.; Jia, H.F.; Shao-Horn, Y. Activity and stability trends of perovskite oxides for oxygen evolution catalysis at neutral ph. *Phys. Chem. Chem. Phys.* **2015**, *17*, 22576–22580. [CrossRef] [PubMed]
24. Kim, B.; Fabbri, E.; Abbott, D.F.; Cheng, X.; Clark, A.H.; Nachtegaal, M.; Borlaf, M.; Castelli, I.E.; Graule, T.; Schmidt, T.J. Functional role of fe-doping in co-based perovskite oxide cata-lysts for oxygen evolution reaction. *J. Am. Chem. Soc.* **2019**. accepted.
25. May, K.J.; Carlton, C.E.; Stoerzinger, K.A.; Risch, M.; Suntivich, J.; Lee, Y.L.; Grimaud, A.; Shao-Horn, Y. Influence of oxygen evolution during water oxidation on the surface of perovskite oxide catalysts. *J. Phys. Chem. Lett.* **2012**, *3*, 3264–3270. [CrossRef]
26. Risch, M.; Grimaud, A.; May, K.J.; Stoerzinger, K.A.; Chen, T.J.; Mansour, A.N.; Shao-Horn, Y. Structural changes of cobalt-based perovskites upon water oxidation investigated by exafs. *J. Phys. Chem. C* **2013**, *117*, 8628–8635. [CrossRef]
27. Fabbri, E.; Nachtegaal, M.; Binninger, T.; Cheng, X.; Kim, B.; Durst, J.; Bozza, F.; Graule, T.; Schäublin, R.; Wiles, L.; et al. Dynamic surface self-reconstruction is the key of highly active perovskite nano-electrocatalysts for water splitting. *Nat. Mater.* **2017**, *16*, 925–932. [CrossRef] [PubMed]
28. Suntivich, J.; Gasteiger, H.A.; Yabuuchi, N.; Nakanishi, H.; Goodenough, J.B.; Shao-Horn, Y. Design principles for oxygen-reduction activity on perovskite oxide catalysts for fuel cells and metal-air batteries. *Nat. Chem.* **2011**, *3*, 546–550. [CrossRef] [PubMed]

29. Burke, M.S.; Kast, M.G.; Trotochaud, L.; Smith, A.M.; Boettcher, S.W. Cobalt-iron (oxy)hydroxide oxygen evolution electrocatalysts: The role of structure and composition on activity, stability, and mechanism. *J. Am. Chem. Soc.* **2015**, *137*, 3638–3648. [CrossRef] [PubMed]
30. Gong, M.; Li, Y.G.; Wang, H.L.; Liang, Y.Y.; Wu, J.Z.; Zhou, J.G.; Wang, J.; Regier, T.; Wei, F.; Dai, H.J. An advanced ni-fe layered double hydroxide electrocatalyst for water oxidation. *J. Am. Chem. Soc.* **2013**, *135*, 8452–8455. [CrossRef] [PubMed]
31. Zou, S.H.; Burke, M.S.; Kast, M.G.; Fan, J.; Danilovic, N.; Boettcher, S.W. Fe (oxy)hydroxide oxygen evolution reaction electrocatalysis: Intrinsic activity and the roles of electrical conductivity, substrate, and dissolution. *Chem. Mater.* **2015**, *27*, 8011–8020. [CrossRef]
32. Trotochaud, L.; Young, S.L.; Ranney, J.K.; Boettcher, S.W. Nickel-iron oxyhydroxide oxygen-evolution electrocatalysts: The role of intentional and incidental iron incorporation. *J. Am. Chem. Soc.* **2014**, *136*, 6744–6753. [CrossRef] [PubMed]
33. Dionigi, F.; Strasser, P. Nife-based (oxy)hydroxide catalysts for oxygen evolution reaction in non-acidic electrolytes. *Adv. Energy Mater.* **2016**, *6*, 1600621. [CrossRef]
34. Chen, J.Y.C.; Dang, L.N.; Liang, H.F.; Bi, W.L.; Gerken, J.B.; Jin, S.; Alp, E.E.; Stahl, S.S. Operando analysis of nife and fe oxyhydroxide electrocatalysts for water oxidation: Detection of fe4+ by mossbauer spectroscopy. *J. Am. Chem. Soc.* **2015**, *137*, 15090–15093. [CrossRef] [PubMed]
35. Gorlin, M.; Chernev, P.; de Araujo, J.F.; Reier, T.; Dresp, S.; Paul, B.; Krahnert, R.; Dau, H.; Strasser, P. Oxygen evolution reaction dynamics, faradaic charge efficiency, and the active metal redox states of ni-fe oxide water splitting electrocatalysts. *J. Am. Chem. Soc.* **2016**, *138*, 5603–5614. [CrossRef] [PubMed]
36. Gorlin, M.; de Araujo, J.F.; Schmies, H.; Bernsmeier, D.; Dresp, S.; Gliech, M.; Jusys, Z.; Chernev, P.; Kraehnert, R.; Dau, H.; et al. Tracking catalyst redox states and reaction dynamics in ni-fe oxyhydroxide oxygen evolution reaction electrocatalysts: The role of catalyst support and electrolyte ph. *J. Am. Chem. Soc.* **2017**, *139*, 2070–2082. [CrossRef] [PubMed]
37. Abbott, D.F.; Fabbri, E.; Borlaf, M.; Bozza, F.; Schäublin, R.; Nachtegaal, M.; Graule, T.; Schmidt, T.J. Operando x-ray absorption investigations into the role of fe in the electrochemical stability and oxygen evolution activity of ni1−xfexoy nanoparticles. *J. Mater. Chem. A* **2018**, *6*, 24534–24549. [CrossRef]
38. Abbott, D.F.; Meier, M.; Meseck, G.R.; Fabbri, E.; Seeger, S.; Schmidt, T.J. Silicone nanofilament-supported mixed nickel-metal oxides for alkaline water electrolysis. *J. Electrochem. Soc.* **2017**, *164*, F203–F208. [CrossRef]
39. Heel, A.; Vital, A.; Holtappels, P.; Graule, T. Flame spray synthesis and characterisation of stabilised zro2 and ceo2 electrolyte nanopowders for sofc applications at intermediate temperatures. *J. Electroceram.* **2009**, *22*, 40–46. [CrossRef]
40. Han, B.H.; Grimaud, A.; Giordano, L.; Hong, W.T.; Diaz-Morales, O.; Yueh-Lin, L.; Hwang, J.; Charles, N.; Stoerzinger, K.A.; Yang, W.L.; et al. Iron-based perovskites for catalyzing oxygen evolution reaction. *J. Phys. Chem. C* **2018**, *122*, 8445–8454. [CrossRef]
41. Binninger, T.; Mohamed, R.; Waltar, K.; Fabbri, E.; Levecque, P.; Kotz, R.; Schmidt, T.J. Thermodynamic explanation of the universal correlation between oxygen evolution activity and corrosion of oxide catalysts. *Sci. Rep.* **2015**, *5*, 12167. [CrossRef] [PubMed]
42. Kim, B.J.; Abbott, D.F.; Cheng, X.; Fabbri, E.; Nachtegaal, M.; Bozza, F.; Castelli, I.E.; Lebedev, D.; Schaublin, R.; Coperet, C.; et al. Unraveling thermodynamics, stability, and oxygen evolution activity of strontium ruthenium perovskite oxide. *ACS Catal.* **2017**, *7*, 3245–3256. [CrossRef]
43. Bick, D.S.; Kindsmuller, A.; Staikov, G.; Gunkel, F.; Muller, D.; Schneller, T.; Waser, R.; Valov, I. Stability and degradation of perovskite electrocatalysts for oxygen evolution reaction. *Electrochim. Acta* **2016**, *218*, 156–162. [CrossRef]
44. Battle, P.D.; Gibb, T.C.; Strange, R. A study of a new incommensurate phase in the system srmn1-xcoxo3-y. *J. Solid State Chem.* **1989**, *81*, 217–229. [CrossRef]
45. Gibb, T.C. Evidence for an unusual phase in the perovskite-related system bacoxmn1-xo3-y from exafs spectroscopy. *J. Mater. Chem.* **1992**, *2*, 387–393. [CrossRef]
46. Gangopadhayay, S.; Inerbaev, T.; Masunov, A.E.; Altilio, D.; Orlovskaya, N. Structural characterization combined with the first principles simulations of barium/strontium cobaltite/ferrite as promising material for solid oxide fuel cells cathodes and high-temperature oxygen permeation membranes. *ACS Appl. Mater. Interfaces* **2009**, *1*, 1512–1519. [CrossRef] [PubMed]

47. Ganopadhyay, S.; Masunov, A.E.; Inerbaev, T.; Mesit, J.; Guha, R.K.; Sleiti, A.K.; Kapat, J.S. Understanding oxygen vacancy migration and clustering in barium strontium cobalt iron oxide. *Solid State Ionics* **2010**, *181*, 1067–1073. [CrossRef]
48. Mueller, D.N.; De Souza, R.A.; Brendt, J.; Samuelis, D.; Martin, M. Oxidation states of the transition metal cations in the highly nonstoichiometric perovskite-type oxide ba0.1sr0.9co0.8fe0.2o3-delta. *J. Mater. Chem.* **2009**, *19*, 1960–1963. [CrossRef]
49. Arnold, M.; Xu, Q.; Tichelaar, F.D.; Feldhoff, A. Local charge disproportion in a high-performance perovskite. *Chem. Mater.* **2009**, *21*, 635–640. [CrossRef]
50. Jun, A.; Kim, J.; Shin, J.; Kim, G. Perovskite as a cathode material: A review of its role in solid-oxide fuel cell technology. *ChemElectroChem* **2016**, *3*, 511–530. [CrossRef]
51. Kuklja, M.M.; Kotomin, E.A.; Merkle, R.; Mastrikov, Y.A.; Maier, J. Combined theoretical and experimental analysis of processes determining cathode performance in solid oxide fuel cells. *Phys. Chem. Chem. Phys.* **2013**, *15*, 5443–5471. [CrossRef] [PubMed]
52. Glazer, A.M. Simple ways of determining perovskite structures. *Acta Cryst. Sect. A* **1975**, *31*, 756–762. [CrossRef]
53. Risch, M. Perovskite electrocatalysts for the oxygen reduction reaction in alkaline media. *Catalysts* **2017**, *7*, 154. [CrossRef]
54. Heel, A.; Holtappels, P.; Hug, P.; Graule, T. Flame spray synthesis of nanoscale la0.65sr0.4co0.2fe0.8o3-delta and ba0.5sr0.5co0.8fe0.2o3-delta as cathode materials for intermediate temperature solid oxide fuel cells. *Fuel Cells* **2010**, *10*, 419–432. [CrossRef]
55. Muller, O.; Nachtegaal, M.; Just, J.; Lutzenkirchen-Hecht, D.; Frahm, R. Quick-exafs setup at the superxas beamline for in situ x-ray absorption spectroscopy with 10 ms time resolution. *J. Synchrotron Radiat.* **2016**, *23*, 260–266. [CrossRef] [PubMed]
56. Ravel, B.; Newville, M. Athena, artemis, hephaestus: Data analysis for x-ray absorption spectroscopy using ifeffit. *J. Synchrotron Radiat.* **2005**, *12*, 537–541. [CrossRef] [PubMed]
57. Delaplane, R.G.; Ibers, J.A.; Ferraro, J.R.; Rush, J.J. Diffraction and spectroscopic studies of cobaltic acid system hcoo2-dcoo2. *J. Chem. Phys.* **1969**, *50*, 1920–1927. [CrossRef]
58. Schmidt, T.J.; Gasteiger, H.A.; Stab, G.D.; Urban, P.M.; Kolb, D.M.; Behm, R.J. Characterization of high-surface area electrocatalysts using a rotating disk electrode configuration. *J. Electrochem. Soc.* **1998**, *145*, 2354–2358. [CrossRef]
59. Binninger, T.; Fabbri, E.; Patru, A.; Garganourakis, M.; Han, J.; Abbott, D.F.; Sereda, O.; Kotz, R.; Menzel, A.; Nachtegaal, M.; et al. Electrochemical flow-cell setup for in situ x-ray investigations i. Cell for saxs and xas at synchrotron facilities. *J. Electrochem. Soc.* **2016**, *163*, H906–H912. [CrossRef]
60. Castelli, I.E.; Huser, F.; Pandey, M.; Li, H.; Thygesen, K.S.; Seger, B.; Jain, A.; Persson, K.A.; Ceder, G.; Jacobsen, K.W. New light-harvesting materials using accurate and efficient bandgap calculations. *Adv. Energy Mater.* **2015**, *5*, 1400915. [CrossRef]
61. The Materials Project. Available online: https://materialsproject.org/ (accessed on 17 November 2017).
62. Johnson, J.W.; Oelkers, E.H.; Helgeson, H.C. Supcrt92 - a software package for calculating the standard molal thermodynamic properties of minerals, gases, aqueous species, and reactions from 1-bar to 5000-bar and 0-degrees-c to 1000-degrees-c. *Comput. Geosci.* **1992**, *18*, 899–947. [CrossRef]
63. Pourbaix, M. Atlas of Electrochemical Equilibria in Aqueous Solutions. Pergamon Press: Oxford, UK, 1966.
64. Larsen, A.H.; Mortensen, J.J.; Blomqvist, J.; Castelli, I.E.; Christensen, R.; Dulak, M.; Friis, J.; Groves, M.N.; Hammer, B.; Hargus, C.; et al. The atomic simulation environment-a python library for working with atoms. *J. Phys. Condens. Mat.* **2017**, *29*, 273002.
65. Persson, K.A.; Waldwick, B.; Lazic, P.; Ceder, G. Prediction of solid-aqueous equilibria: Scheme to combine first-principles calculations of solids with experimental aqueous states. *Phys. Rev. B* **2012**, *85*. [CrossRef]
66. Castelli, I.E.; Thygesen, K.S.; Jacobsen, K.W. Calculated optical absorption of different perovskite phases. *J. Mater. Chem. A* **2015**, *3*, 12343–12349. [CrossRef]
67. Giannozzi, P.; Baroni, S.; Bonini, N.; Calandra, M.; Car, R.; Cavazzoni, C.; Ceresoli, D.; Chiarotti, G.L.; Cococcioni, M.; Dabo, I.; et al. Quantum espresso: A modular and open-source software project for quantum simulations of materials. *J. Phys. Condens. Mat.* **2009**, *21*, 395502. [CrossRef] [PubMed]

68. Perdew, J.P.; Ruzsinszky, A.; Csonka, G.I.; Vydrov, O.A.; Scuseria, G.E.; Constantin, L.A.; Zhou, X.L.; Burke, K. Restoring the density-gradient expansion for exchange in solids and surfaces. *Phys. Rev. Lett.* **2008**, *100*. [CrossRef] [PubMed]
69. Standard Solid State Pseudopotential Library. Available online: http://materialscloud.org/sssp (accessed on 17 November 2017).

© 2019 by the authors. Licensee MDPI, Basel, Switzerland. This article is an open access article distributed under the terms and conditions of the Creative Commons Attribution (CC BY) license (http://creativecommons.org/licenses/by/4.0/).

Article

K-Modulated Co Nanoparticles Trapped in La-Ga-O as Superior Catalysts for Higher Alcohols Synthesis from Syngas

Shaoxia Guo [1,2], Guilong Liu [3,4,*], Tong Han [5], Ziyang Zhang [1,2] and Yuan Liu [1,2,*]

1. Tianjin Key Laboratory of Applied Catalysis Science and Technology, School of Chemical Engineering, Tianjin University, Tianjin 300350, China; guosx@tju.edu.cn (S.G.); ziyang_zhang@tju.edu.cn (Z.Z.)
2. Collaborative Innovation Center of Chemical Science and Engineering (Tianjin), Tianjin 300072, China
3. College of Chemistry and Chemical Engineering, Luoyang Normal University, Luoyang 471934, China
4. Key Laboratory of Function-oriented Porous Materials of Henan Province, Luoyang Normal University, Luoyang 471934, China
5. Department of Material Science and Engineering, KTH Royal Institute of Technology, 11428 Stockholm, Sweden; tongh@kth.se
* Correspondence: glliu@tju.edu.cn (G.L.); yuanliu@tju.edu.cn (Y.L.); Tel.: +86-0379-68618321 (G.L.); +86-022-87401675 (Y.L.)

Received: 11 February 2019; Accepted: 22 February 2019; Published: 27 February 2019

Abstract: Owing to the outstanding catalytic performance for higher alcohol synthesis, Ga-Co catalysts have attracted much attention. In view of their unsatisfactory stability and alcohol selectivity, herein, K-modulated Co nanoparticles trapped in La-Ga-O catalysts were prepared by the reduction of $La_{1-x}K_xCo_{0.65}Ga_{0.35}O_3$ perovskite precursor. Benefiting from the atomic dispersion of all the elements in the precursor, during the reduction of $La_{1-x}K_xCo_{0.65}Ga_{0.35}O_3$, Co nanoparticles could be confined into the K-modified La-Ga-O composite oxides, and the confinement of La-Ga-O could improve the anti-sintering performance of Co nanoparticles. In addition, the addition of K modulated parts of La-Ga-O into La_2O_3, which ameliorated the anti-carbon deposition performance. Finally, the addition of K increased the dispersion of cobalt and provided more electron donors to metallic Co, resulting in a high activity and superior selectivity to higher alcohols. Benefiting from the above characteristics, the catalyst possesses excellent activity, good selectivity, and superior stability.

Keywords: perovskite-type oxide (PTO); cobalt; gallium; potassium; higher alcohols; syngas

1. Introduction

Due to their sufficient combustion and release of less harmful substances during combustion, higher alcohols with 2–6 carbon atoms are regarded as a kind of clean energy [1]. In addition, due to the high octane number, higher alcohols can also be used as a high-quality fuel additive. After separation, a series of basic chemicals with very high economic value, such as ethanol, propanol, and butanol, can be obtained [2–4]. Currently, ethanol is mainly produced by fermentation and ethylene hydration, while other alcohols are refined from petroleum. Obviously, in the long run, the above synthesis routes for higher alcohol would be restricted by increasingly depleted petroleum and food [5]. Recently, the synthesis of higher alcohols from syngas has attracted much attention, while this process is usually restricted by the low selectivity to higher alcohols and the poor stability of the catalyst.

Nowadays, four kinds of catalysts for higher alcohol synthesis (HAS) from syngas have been reported. Among them, the Rh-based catalysts show good activity and superior selectivity to ethanol, while the high price of Rh limits its industrial applications [6,7]. The harsh reaction conditions usually restrict large-scale applications for Mo-based catalysts [8,9]. For modified methanol synthesis catalysts, the main product is still methanol [10,11]. Fortunately, modified Fischer–Tropsch catalysts, mainly the modified Co and modified Fe catalysts, exhibit good activity and high selectivity for HAS at milder reaction conditions. However, the modified Fe catalysts are more beneficial to the water gas shift reaction (WGSR), generating lots of CO_2; and the typical Cu modified Co catalysts usually show poor stability because of the phase separation of cobalt and copper [12–14]. Therefore, it has become an important issue for researchers to explore new catalysts for HAS from syngas with better catalytic performance.

Recently, Ga-modified Co catalysts were reported and showed excellent catalytic performance for higher alcohol synthesis [15–17]. He et al. prepared a series of Co-Ga catalysts by using Co-Ga-LDHs (layered double hydroxides) and found that Ga was beneficial to the non-dissociative adsorption of CO [15,16]. Gao et al. reported that gallium oxide can reduce the reduction degree of CoO and generate some Co^{2+} in the reduced catalysts, which act as non-dissociative CO adsorption sites for HAS, resulting in the high selectivity to alcohols for the Ga-Co/AC catalyst [17]. While the stability of Co-Ga catalysts should be further improved.

Here, considering the good activity and high selectivity on Co-Ga catalysts, K doped Co-Ga catalysts are explored by the reduction of $La_{1-x}K_xCo_{0.65}Ga_{0.35}O_3$. The results show that the addition of K modulates the composition of La-Ga-O, enhances the dispersion of Co, and adjusts the electronic structure of Co, and as a result the catalysts possess excellent catalytic performance. Typically, an outstanding selectivity of 43.6% to the higher alcohols, and a stable catalytic performance during the 200 h reaction can be obtained.

2. Results and Discussion

2.1. X-ray Powder Diffraction (XRPD)

The X-ray powder diffraction (XRPD) patterns of the three catalysts for LCG ($LaCo_{0.65}Ga_{0.35}O_3$) and LKCG-x ($La_{1-x}K_xCo_{0.65}Ga_{0.35}O_3$, x = 0.1 and 0.2, where x is the K content in perovskite) (see 3.2 catalysis synthesis) are shown in Figure 1a. The diffraction peaks at 2θ = 23.2, 33.3, 40.6, 47.4 and 58.8° are attributed to the characteristic diffraction peaks of perovskite-type oxide (PTO). For LKCG-0.1 catalyst, with the addition of K ions into LCG, the perovskite diffraction peaks move to lower 2θ values (seen from the illustration of Figure 1a), for that the ion radius of K^+ (0.155 nm) is larger than that of La^{3+} (0.136 nm) [18]. The existence of perovskite structure after calcination is beneficial to the interaction and the even dispersion of all the elements.

For the LKCG-0.2, a new Co_3O_4 diffraction peak in Figure 1a can be seen. Since the amount of K entering the perovskite is limited, when the K doping amount is more than 0.1, part of potassium cannot incorporate into the perovskite structure and cover the surface of the catalyst in the form of oxide [18]. The presence of K_2O disrupted the dispersion of elements in the catalyst precursor, resulting in the formation of Co_3O_4. It is worth noting that a part of $LaCo_yGa_{1-y}O_3$ and $La_{1-z}K_zCo_{1-m}Ga_mO_3$ should also exist accompanied by the formation of Co_3O_4.

Meanwhile, Ga-containing oxides among the three samples cannot be detected, indicating that Ga entered into the structure of perovskite. The uniformly dispersed Co and Ga ions in the LKCG-0.1 catalyst are advantageous for the synergism between them, favoring the catalytic performance.

The XRPD profiles of three catalysts reduced at 750 °C (see 3.4 Catalysts' Performance) are presented in Figure 1b. As for LCG, the perovskite structure disappears and transfers to Co and $La_4Ga_2O_9$. As for the reduction of LKCG-0.1, phases of La_2O_3, Co, $LaGaO_3$, and a small amount of $La_4Ga_2O_9$ can be observed. The existence of the characteristic diffraction peak of $LaGaO_3$ and La_2O_3 indicated that the adding of K weakened the interaction between lanthanum and gallium. In other words, the addition of K in the perovskite modulated the composition of La-Ga-O, resulting in the change of La-Ga-O from $La_4Ga_2O_9$ to $LaGaO_3$ and La_2O_3.

Based on the above discussion in Figure 1a, parts of K cannot be doped into perovskite, resulting co-exist of $LaCo_yGa_{1-y}O_3$ and $La_{1-z}K_zCo_{1-m}Ga_mO_3$ in LKCG-0.2. During reduction,

La$_{1-z}$K$_z$Co$_{1-m}$Ga$_m$O$_3$ would be reduced into LaGaO$_3$ while LaCo$_y$Ga$_{1-y}$O$_3$ to La$_4$Ga$_2$O$_9$, as a result, the LKCG-0.2 are reduced to Co/LaGaO$_3$-La$_4$Ga$_2$O$_9$, as can be seen in Figure 1b.

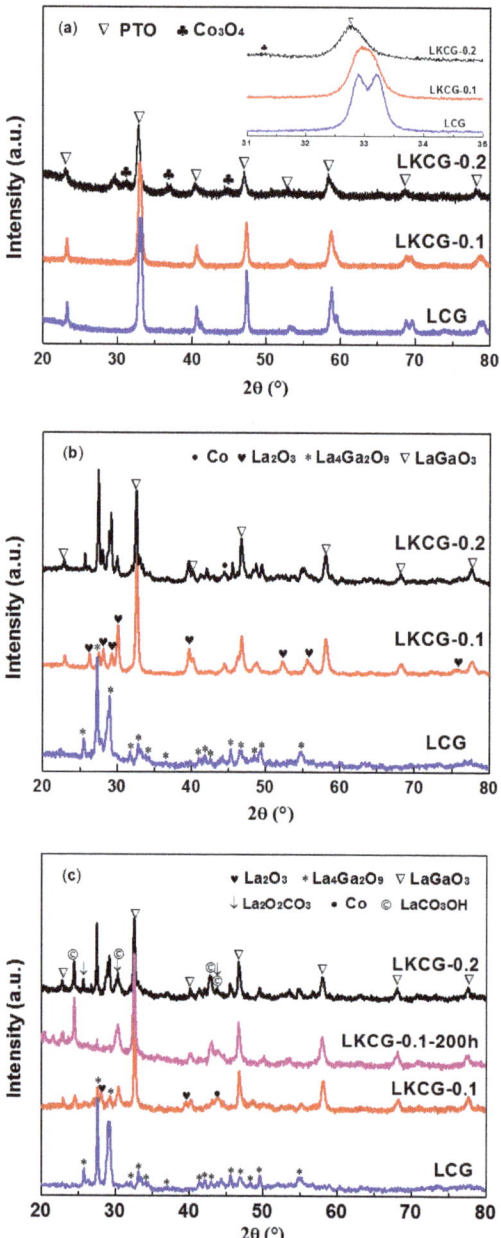

Figure 1. X-ray powder diffraction (XRPD) patterns of catalysts after (**a**) calcination, (**b**) reduction and (**c**) reaction.

Figure 1c shows the XRPD profiles of the three catalysts after reaction and LKCG-0.1 after 200 h reaction. After reaction, part of La_2O_3 transferred to $LaCO_3OH$ and $La_2O_2CO_3$, for that La_2O_3 and CO_2 can react to generate $La_2O_2CO_3$, and the further reaction between $La_2O_2CO_3$, H_2O and CO_2 can generate $LaCO_3OH$ [19,20]. Since XRPD in this work was carried out ex situ, the catalysts containing $La_2O_2CO_3$ could readily absorb H_2O and CO_2 in air, and then $LaCO_3OH$ formed. The co-existence of La_2O_3 and $La_2O_2CO_3$ in the catalysts after reaction illustrated the feasibility of reaction of CO_2 + $La_2O_3 \rightarrow La_2O_2CO_3 \xrightarrow{C} 2CO + La_2O_3$, which can help the catalysts eliminate carbon.

For the LCG catalyst after reaction, it should be noted that the catalyst was still $Co/La_4Ga_2O_9$, which is the same as that after reduction. $La_2O_2CO_3$ and La_2O_3 cannot be detected, indicating the above reaction of eliminating carbon deposition may be hard to occur due to a strong interaction existing between lanthanum and gallium.

It should be noted that no Co_2C was observed in all the used samples, suggesting that the existence of gallium can prevent the formation of Co_2C and stabilize the catalyst composition in the process of reaction. This is in accordance with the literature, which illustrated that the existence of gallium could improve the catalyst's stability [21].

2.2. Temperature-Programmed Reduction (TPR)

Figure 2 and Table 1 illustrates the temperature-programmed reduction (TPR) and the hydrogen consumptions values of LCG and LKCG-x (x = 0.1 and 0.2) catalysts. Seen from the Figure 2, all the three catalysts contain two major hydrogen-consuming peaks, one at 400–500 °C and the other at 600–800 °C. According to the literature, the H_2-consuming peak of $LaCoO_3$ at 500 °C can be attributed to $Co^{3+} \rightarrow Co^{2+}$, while that above 500 °C to $Co^{2+} \rightarrow Co^0$ [22–25]. Herein, seen from Table 1, the area ratio of the low temperature peak to the high temperature peak (T_L/T_H) is 1/2. Therefore, we believe that the peak of 400–500 °C in all the three catalysts is classified to $Co^{3+} \rightarrow Co^{2+}$, and the peak of 600–800 °C can be assigned to $Co^{2+} \rightarrow Co^0$. At the same time, the similar theoretical and experimental H_2-consumption also confirmed the attribution of the above reduction peak.

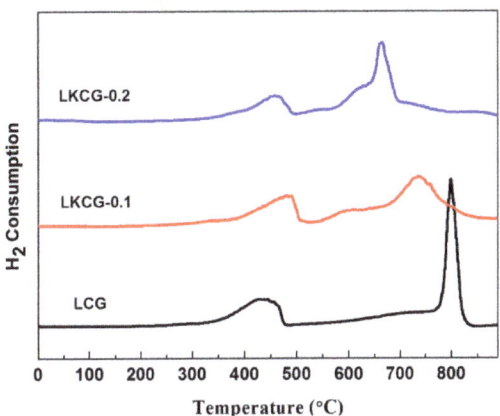

Figure 2. The temperature-programmed reduction (TPR) curves of catalysts after calcination.

Compared to the TPR results of $LaCoO_3$ in the literature, the H_2-consuming peak of $Co^{2+} \rightarrow Co^0$ migrates to higher temperatures in the LCG catalyst, suggesting that the existence of Ga in $LaCoO_3$ will restrain the conversion of Co^{2+} to Co^0 and a strong effect between Co and Ga exists for LCG [16].

Seen from Figure 2, with the addition of K ions in LCG, the H_2-consuming peaks at around 750 °C moved to lower temperature, illustrating the addition of K promotes the reduction of $Co^{2+} \rightarrow Co^0$. While for LKCG-0.2, a shoulder peak around 620 °C appeared. According to XRPD, the catalyst contains Co_3O_4. Therefore, this small shoulder can be attributed to the reduction of $Co^{2+} \rightarrow Co^0$ in

Co$_3$O$_4$. The presence of the small shoulder also confirms the XRPD results. For the other samples, no shoulders can be observed, which indicates that all the Co ions have entered into the crystal lattice of the perovskite.

Table 1. Theoretical and experimental H$_2$ consumption value of the LKCG-x (x = 0.1 and 0.2) and LCG catalysts.

Catalysts	Experimental Measure [a,b]		Theoretical Measure	
	T$_L$	T$_H$	Co^{3+} → Co^{2+}	Co^{2+} → Co
LCG	0.064	0.132	0.065	0.130
LKCG-0.1	0.067	0.133	0.068	0.135
LKCG-0.2	0.069	0.136	0.070	0.140

[a] Calculated from the TPR results. [b] The unit of H$_2$ consumption value is mmol H$_2$/50 mg. T$_L$ and T$_H$ represent low and high temperature, respectively.

2.3. N$_2$ Adsorption and Desorption Curves

Figure 3 and Table 2 show the N$_2$ physical adsorption curves and the physical properties of the investigated catalysts. Seen from the Figure 3, the curves are typical type II isotherms accompanying with a H$_3$ type hysteresis loop, which indicates that mesoporous exits in the catalysts. The presence of mesoporous can also be seen from the pore size distribution curves.

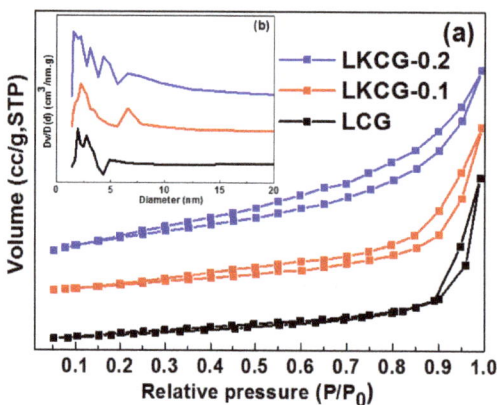

Figure 3. (a) The N$_2$ physical adsorption curves and (b) Barrett–Joyner–Halenda (BJH) pore size distribution of catalysts.

Table 2. The physical properties of all the investigated catalysts.

Catalysts	S$_{BET}$ (m^2 g^{-1})	Pore Size (nm)	V$_{BJH}$ (cm^3 g^{-1})	Crystal Size (nm) [a]		D$_{Co}$ (%) [e]	d$_{Co}$ (nm) [f]
				PTO	Co [b]		
LCG	8.6	14.9	0.03	17.4	15.9	7.3 [b] (6.1) [c]	13.2 [b] (15.7) [c]
LKCG-0.1	11.7	12.6	0.04	16.2	8.3	13.7 [b] (11.4) [c] (9.9) [d]	7.0 [b] (8.4) [c] (9.7) [d]
LKCG-0.2	15.0	9.8	0.04	13.4	7.4	15.5 [b] (12.6) [c]	6.2 [b] (7.6) [c]

[a] Calculated from X-ray diffraction results with the Scherrer equation. [b] The reduced catalysts. [c] The catalysts after reaction. [d] The 200 h stability test. [e] The degree of dispersion. [f] The crystal size calculated by hydrogen temperature-programmed desorption (H$_2$-TPD).

For LKCG-x (x = 0.1 and 0.2) catalysts, with the addition of K, the hysteresis loop increases and moves to lower P/P$_0$, illustrating the investigated catalysts have bigger BET surface area. The larger surface area results in the higher dispersion of metal cobalt nanoparticles (see in Table 2).

2.4. X-ray Photoelectron Spectroscopy (XPS)

Figure 4 and Table 3 summarized the binding energies (BEs) and the X-ray photoelectron spectroscopy (XPS) profiles of La 3d, Co 2p, and Ga 3d for the reduced LCG and LKCG-0.1 catalysts. All XPS profiles showed almost similar peak patterns except the different value of binding energies of each element. The binding energy of Co in the both samples is similar to that of metal cobalt, illustrating that Co exists in the form of Co^0 in the catalyst, which is the same as the XRPD results.

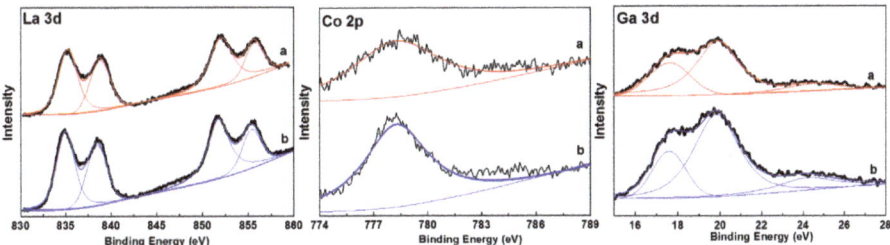

Figure 4. X-ray photoelectron spectroscopy (XPS) profiles of La 3d, Co 2p and Ga 3d of the reduced (**a**) LCG and (**b**) LKCG-0.1.

Table 3. The binding energies of the reduced LCG and LKCG-0.1 catalysts.

Catalysts	La $3d_{5/2}$		Co $2p_{3/2}$	Ga $3d_{5/2}$	
LCG	835.2	838.6	778.3	17.6	19.7
LKCG-0.1	834.9	838.4	778.1	17.7	19.8

According to the reported literature, the binding energies of La $3d_{5/2}$ are 834.4 and 837.8 eV for pure La_2O_3, 834.7 eV and 838.1 eV for $LaGaO_3$ [26–28]. Herein, the binding energies of La $3d_{5/2}$ for LCG are 835.2 eV and 838.6 eV, which is larger than that of pure La_2O_3 and $LaGaO_3$. The binding energies of Ga $3d_{5/2}$ are 17.6 and 19.7 eV, which is also a little larger than that of $LaGaO_3$ at 17.4 and 19.4 eV [28]. The binding energies of Co $2p_{3/2}$ is 778.3 eV, which is less than 778.5 eV for metal cobalt [29]. The higher binding energies of La $3d_{5/2}$ and Ga $3d_{5/2}$, and the lower binding energies of Co $2p_{3/2}$ illustrate that an interaction among La, Ga, and Co existed. At the same time, La and Ga could donate elector to Co.

Compared to the reduced LCG catalyst, it was found that the binding energy of La in LKCG-0.1 decreased, suggesting that the doping of K modulated the interaction between La and Ga, which agrees with the above XRPD results. In addition, the binding energy of Co is lower, which means that the K could donate electron to Co. The enhanced electron for Co is beneficial for the selectivity to higher alcohols.

2.5. Transmission Electron Microscopy (TEM)

Figure 5a–h shows the transmission electron microscopy (TEM) images, the line scans profiles, the energy dispersive spectrometer (EDS) mapping scans image, and the elements distribution of the reduced LKCG-0.1 catalyst. In Figure 5a, 5–11 nm Co nanoparticles are uniform dispersed in the reduced LKCG-0.1 catalyst even after 750 °C high temperature reduction. In Figure 5b, the lattice spacing of [112] and [011] planes for La_2O_3, [200] and [111] planes for metal Co, [220] planes for $La_4Ga_2O_9$ and $LaGaO_3$ can be clearly seen. That is to say, the composition of the reduced LKCG-0.1 catalyst is Co/La_2O_3-$La_4Ga_2O_9$-$LaGaO_3$, which is consistent with the XRPD. In addition, as seen from Figure 5b, the metal cobalt nanoparticles are encircled and located between La-Ga-O oxides. And this confinement effect result in the highly dispersion of Co nanoparticles in Figure 5a.

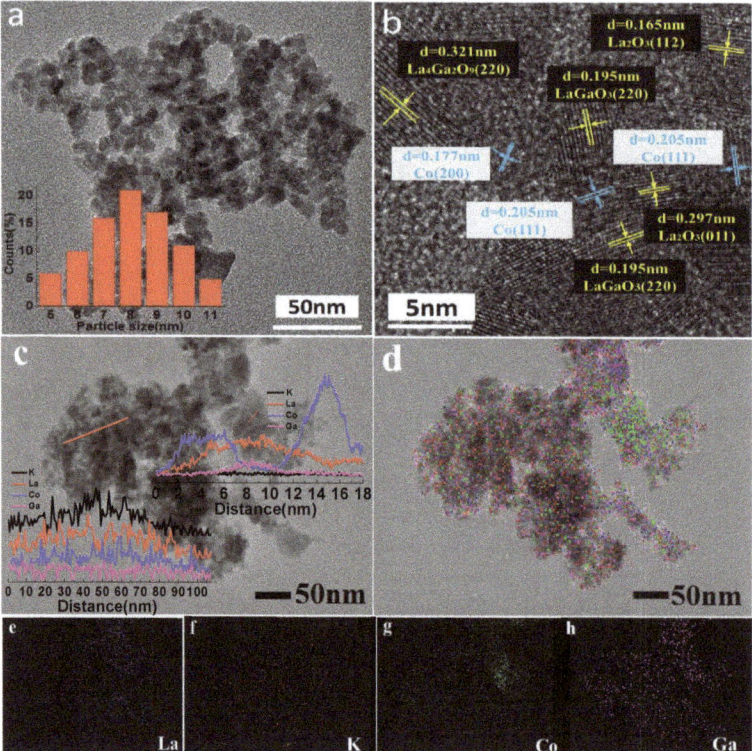

Figure 5. Transmission electron microscopy (TEM) images (**a**–**c**), line scanning profiles, energy dispersive spectrometer (EDS) mapping image (**d**), and the elements distribution of La (**e**), K (**f**), Co (**g**), and Ga (**h**) for the reduced LKCG-0.1.

Figure 5c–h exhibits the line scans profiles and the corresponding element distribution of the reduced LKCG-0.1 catalyst. The red lines represent the scanning routes in Figure 5c. Seen from the illustration in Figure 5c, the La and Ga have the same change trend, indicating the formation of La-Ga-O. In addition, there is no La and Ga where Co appears, illustrating metal cobalt nanoparticles are highly dispersed and located between the La-Ga-O oxides, which is in accordance with the XRPD and Figure 5b.

Figure 6a–h displays the TEM images, the line scanning profiles, the EDS mapping image, and the corresponding elements distribution of LKCG-0.1 after 200 h stability tests. As can be seen from Figure 6a, Co is still located between La-Ga-O oxide and the average crystal size of the Co nanoparticles is 9.5 nm, indicating that the sintering of the catalyst is not obvious, which is consistent with the results in Table 2. Seen from the Figure 6b, the lattice spacing of 0.205 and 0.177 nm are assigned to parameters of the [111] and [200] planes of Co; the lattice spacing of 0.228 nm, 0.276 nm, 0.306 nm and 0.294 nm corresponds to the [012], [200], [023], and [103] planes for La_2O_3, $LaGaO_3$, $La_4Ga_2O_9$ and $La_2O_2CO_3$, respectively. That is to say, the component of the LKCG-0.1 catalyst after 200 h reaction is Co/La_2O_3-$La_4Ga_2O_9$-$LaGaO_3$-$La_2O_2CO_3$, which is in accordance with the XRPD results.

2.6. CO Hydrogenation Performance

Table 4 lists the catalytic performance of all the investigated catalysts. The carbon monoxide hydrogenation performance of the different molar ratio of Co/Ga for $LaCo_yGa_{1-y}O_3$ catalyst were explored in our lab, and the results revealed the optimum molar ratio is 0.65/0.35 [30]. Therefore, the molar ratio of Co/Ga of all the investigated samples was fixed at 0.65/0.35.

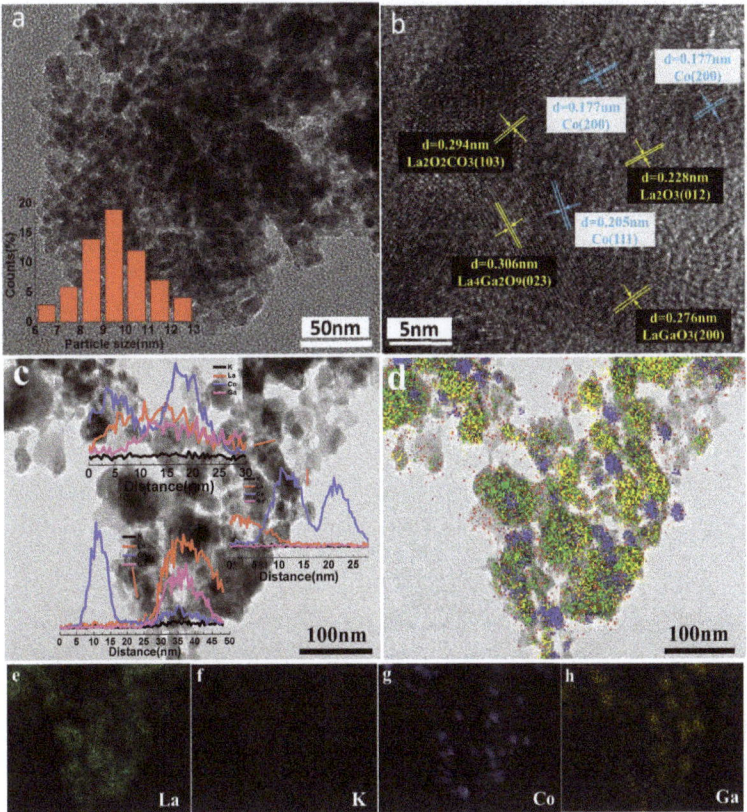

Figure 6. TEM images (a–c), line scanning profiles, EDS mapping image (d), and the elements distribution of La (e), K (f), Co (g), and Ga (h) for LKCG-0.1 after 200 h stability tests.

Table 4. CO hydrogenation performance of the investigated catalysts.

Catalysts	X_{CO} [a] (%)	S_{CO_2} [b] (%)	S_{ROH} [c] (%)	Selectivity to Hydrocarbon (%)				Distribution of Alcohols (%)			
				C_1	C_2	C_3	C_{4+}	C_1	C_2	C_3	C_{4+}
LCG	4.1	1.6	27.2	45.1	10.5	10.8	4.9	36.3	45.2	1.6	16.9
LKCG-0.1	13.2	5.8	43.6	24.7	9.2	11.9	4.8	30.9	58.1	3.2	7.8
LKCG-0.2	11.7	6.6	40.0	26.3	11.5	11.7	3.9	27.8	59.1	3.9	9.2

Reaction conditions: P = 4 MPa, $H_2/CO/N_2$ = 8/4/1, T = 290 °C, GHSV = 6000 mL $g_{cat}^{-1} \cdot h^{-1}$. [a] X_{CO} is CO conversion. [b] S_{CO_2} is the selectivity to CO_2. [c] S_{ROH} is the selectivity to the total alcohols. i in C_i is number of carbon atoms for the carbon-contained products. C_{4+} represents the carbon-contained products with 4 or more carbon atoms.

As for LCG, seen from the above TEM and XRPD results, the composition after reduction is the same with that after reaction, and all are Co/La$_4$Ga$_2$O$_9$. In addition, the smaller BET surface area for LCG makes the Co nanoparticle severely sintered and unevenly dispersed (see in Table 2), thus resulting in a poor activity. At the same time, the larger Co particle sizes and the strong effects between Ga and La in the catalyst are also detrimental to the generation of the Co-Ga interfaces. The interfaces of Co and Ga were usually considered to be the active sites for HAS [16], while metal cobalt was the active sites of hydrocarbon generation [14]. Therefore, the LCG catalyst has the highest hydrocarbon selectivity among all the samples.

However, as for LKCG-0.1 catalyst, the main composition after reduction is Co/K_2O-La_2O_3-LaGaO$_3$ and the main composition after reaction is Co/K_2O-La_2O_3-$La_2O_2CO_3$-LaGaO$_3$ (seen from the XRPD results). It is known from Table 4 that the catalyst with the optimal catalytic performance is LKCG-0.1. Since they have a larger specific surface area, cobalt nanoparticles are highly dispersed on the catalyst surface, which can be seen in Figure 5 and Table 2. In addition, since all the elements located in the lattice of perovskite, smaller size and uniformly dispersed cobalt nanoparticles are also conducive to generating more Co-Ga interfaces. In the process of reaction, cobalt exists in the form of Co^0. The close contact of cobalt with the Co-Ga interface at the atomic level is beneficial to the synergistic effect of the catalyst, and thus the LKCG-0.1 catalyst exhibits the best catalytic performance. In addition, the electron donating effect of K can promote the increase of the selectivity of higher alcohols [31–34].

For the LKCG-0.2 catalyst, with the increasing of K content, the catalytic activity decreased, for that the addition of K can make part of Co outside the perovskite structure, resulting in a non-uniform dispersion of Co and Ga. A relatively lower activity and selectivity is observed in Table 4.

Other catalysts with outstanding performance reported in the literature are revealed in Table 5 [17,35–38]. By comparison, the activity of LKCG-0.1 catalyst in this work is not the optimal, but it may be one of the good catalysts in general considering relatively lower reaction temperatures and higher alcohol selectivity for HAS.

Table 5. Performance of CO hydrogenation reported in the literature.

Catalysts	Temperature (°C)	H_2/CO [a]	Pressure (MPa)	GHSV (h^{-1})	X_{CO} (%)	S_{ROH} (%)	EtOH (%) [b]	C_{2+}OH [c] (%)	Ref.
LaCo$_{0.7}$Cu$_{0.3}$O$_3$	300	2	6.9	15000	16.0	38.1	37.0	53.8	[37]
Cu-Co/La$_2$O$_3$-SiO$_2$	330	2	3	3900	32.1	39.5	47.5	66.1	[36]
Cu-Co/Al$_2$O$_3$	250	2	2	1800	23.2	23.3	-	79.3	[35]
Co$_3$Cu$_1$-11%CNT	300	2	5	7000	26.5	49.8	-	69.9	[38]
15Co-2.5Ga/AC	220	2	3	4000	13.1	30.3	-	24.5	[17]
LKCG-0.1	290	2	4	6000	13.2	43.6	58.1	69.1	This work

[a] Molar ratio. [b] The ethanol's mass fraction in all alcohols. [c] The mass fraction of the higher alcohols in all alcohols.

Figure 7 presents the carbon monoxide hydrogenation performance for 200-h stability tests of LKCG-0.1 catalyst. Seen from Figure 7, the alcohol's selectivity and CO conversion still maintained stability, which are remained at 19.8% and 41.8%, and the higher alcohols in all alcohols stabilized at 72.8%. The outstanding stability can be owned to the uniform dispersion of the active sites, stable catalyst structure, good sintering resistance, and more Co-Ga interfaces.

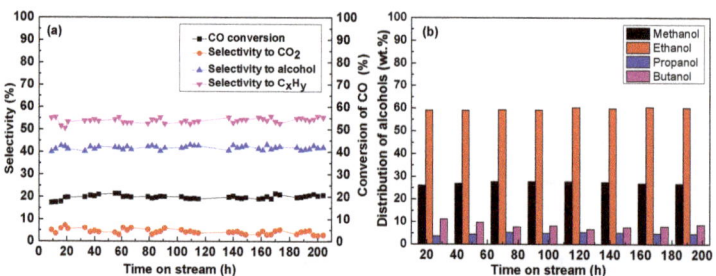

Figure 7. (a) Stability performance and (b) alcohols distributions (seen from the illustration) of the LKCG-0.1 catalyst after 200 h reaction at T = 290 °C, P = 4 MPa, GHSV = 6000 mL g_{cat}^{-1} h^{-1}, and H_2/CO/N_2 = 8/4/1.

2.7. Thermo-Gravimetry (TG)

Figure 8 exhibited the TG curves of the reduced and used LCG and LKCG-x (x = 0.1 and 0.2) catalysts and the corresponding differential thermal gravity (DTG) curves of the used catalysts. In Figure 8, the weight of the three reduced samples increases in the temperature range of 200–320 °C, which is attributed to the oxidation of metal cobalt nanoparticles on the surface of the catalysts. Therefore, the TG profiles of the catalysts after reduction was severed as a datum to explore the carbon deposition amount of the catalysts after reaction. In Figure 8b, two exothermic peaks can be seen for all the samples. The peak located at 300–600 °C can be attributed to amorphous carbon; while the other peak at 600–800 °C to the graphitized carbon [39]. Seen from Figure 8b, the incorporation of K can significantly reduce the total amount and formation rate of amorphous carbon, and for that K can modulate the composition of the catalysts and produce amount of La_2O_3, which is beneficial to the coke elimination.

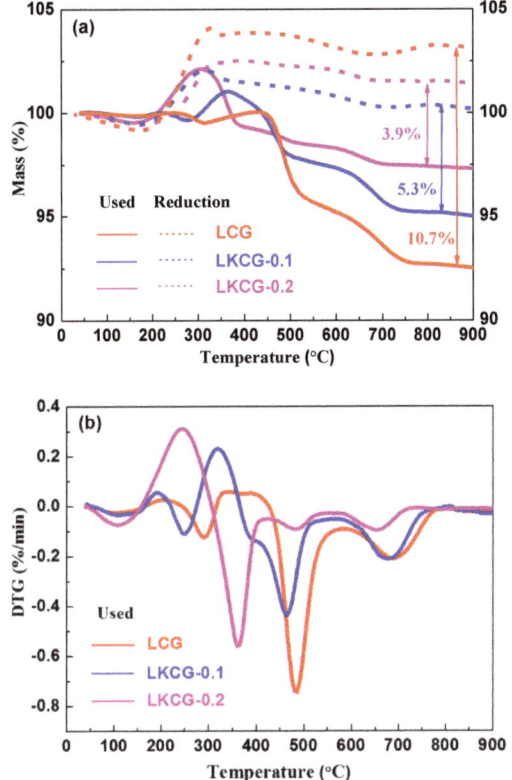

Figure 8. (a) Thermo-gravimetric (TG) curves of the reduced and used catalysts and (b) differential thermal gravity (DTG) curves of the used catalysts.

Seen from the Figure 8, the carbon deposition amounts of LKCG-0.2, LKCG-0.1 and LCG after reaction are 3.9%, 5.3% and 10.7%, respectively; in other words, the addition of K significantly relieves the formation of carbon deposition.

For LKCG-x (x = 0.1 and 0.2) catalysts, the containing carbon content of the two catalysts is almost similar. Compared to LCG catalyst, the adding of K leads to a decrease of carbon deposition. According to the above XRPD and XPS results for the reduced LKCG-x catalysts, the doping of K can modulate the composition of La-Ga-O, generating more La_2O_3, which has a better carbon-depleting effect.

Herein, the formula of $CO_2 + La_2O_3 \rightarrow La_2O_2CO_3 \xrightarrow{C} 2CO + La_2O_3$ is used to illustrate the process of removing carbon deposits during the reaction, indicating that more La_2O_3 favors the anti-carbon effect of catalysts [19,40–42]. In addition, seen from the used XRPD pattern for LKCG-x catalysts, both La_2O_3 and $La_2O_2CO_3$ can also be detected, explaining that the above mechanism is correct. Therefore, the K-doped catalyst exhibits the best anti-carbon deposition performance.

For the LCG catalyst, seen from the used XRPD result, no La_2O_3 and $La_2O_2CO_3$ can be detected. Meanwhile, there is a strong effects between La and Ga, which is not conducive to eliminating carbon deposited, and thus the carbon content of the LCG catalyst is largest.

3. Materials and Methods

3.1. Material

Lanthanum (III) nitrate hexahydrate, citrate acid, cobalt (II) nitrate hexahydrate, potassium nitrate, and glycol were bought from Shanghai Aladdin. Gallium (III) nitrate nonahydrate was purchased from Beijing HWRK Chem. All of above materials were used without further purification.

3.2. Catalysts' Synthesis

The citrate complexation method was used to prepare the K-doping catalysts [43]. Firstly, a solution with La, K, Co, Ga nitrates and citric acid, glycol are mixed by using deionized water, in which the molar ratio of lanthanum:potassium:cobalt:gallium:citric acid:glycol is 1−x:x:0.65:0.35:2.4:0.48. Secondly, the above prepared mixed solution was continuously stirred to gel at 80 °C, and then dried overnight at 120 °C. Finally, the powder catalysts were calcined at 350 °C for 2 h and 700 °C for 5 h (with a heating speed of 2 °C min^{-1}), respectively. The obtained $La_{1-x}K_xCo_{0.65}Ga_{0.35}O_3$ were labeled as LKCG-x (x = 0.1 or 0.2, which is the content of K in perovskite).

For comparison, $LaCo_{0.65}Ga_{0.35}O_3$ without K doping was prepared according to the above method, and the sample was labeled as LCG.

3.3. Catalysts' Characterization

XRPD patterns were performed at the speed of 8 °C min^{-1} between 20° and 80° (2θ). TPR were performed between 30° and 900 °C in 5% H_2/Ar (50 mL min^{-1}) at the heating rate 10 °C min^{-1}. TEM images, the EDS line scans and the corresponding element mapping scans analysis were performed to observe the component structure of the catalysts. The N_2 physical adsorption curve were tested to calculate the BET surface areas, the pore size distributions, and the pore volumes of the catalysts. XPS profiles were used to analyze the binding energy of elements, and the binding energy (BE) of C 1s was 284.6 eV. TG was performed between 30 °C and 900 °C in air at the heating speed of 10 °C min^{-1}. H_2-TPD was used to analyze the dispersion degree and size of metal cobalt. The formula of d (Co^0) = 96/D was used to calculate the size of metal cobalt, and the dispersion of metal cobalt was conducted by using the method reported in literatures [44,45].

3.4. Catalysts' Performance

The CO hydrogenation performance of catalysts were tested in a fixed-bed reactor; 0.4 g samples and 0.4 g quartz sand in 40–60 mesh were mixed and used for the HAS reaction. Temperature, pressure and gas hourly space velocity were set to 290 °C, 4 MPa and 6000 mL·g_{cat}^{-1}·h^{-1}, respectively. The molar ratio of H_2/CO is 2/1, and the internal standard gas of reaction was N_2. Before reaction, the catalysts precursors were reduced at 750 °C and remained 3 h in pure H_2 atmosphere. A TCD detector was used for analyzing H_2, N_2, CO, CO_2, and CH_4; the hydrocarbons and alcohols were detected by using a FID detector. The following formula is used to calculate CO conversion (X_{CO}), the selectivity to product (S_i), and the weight fraction to product (W_i):

$$X_{CO} = \frac{CO_{in} - CO_{out}}{CO_{in}} \times 100\% \tag{1}$$

$$S_i = \frac{nC_i}{\sum nC_i} \times 100\% \tag{2}$$

$$W_i = \frac{m_i}{\sum m_i} \times 100\% \tag{3}$$

CO_{in} and CO_{out} represent the moles fraction of CO in the inlet and outlet gases, respectively; n is the carbon atoms number, and C_i is the mole fractions of carbon-containing products. m_i is the weight of alcohol.

4. Conclusions

$La_{0.9}K_{0.1}Co_{0.65}Ga_{0.35}O_3$ with a perovskite structure was prepared by using the citrate complexation method. Due to all the components being derived from the perovskite structure, after reduction Co is firmly confined to the K-modified La-Ga-O composite oxides, resulting in excellent anti-sintering performance. The addition of K can modulate the composition of La-Ga-O, forming more La_2O_3, favoring the improvement of anti-carbon deposition performance. In addition, the addition of K also increases the dispersion of cobalt, which can generate a greater Co-Ga interface. What is more important is that the doping of K can provide more electron donors for metallic Co, which enhances the selectivity to higher alcohols. Therefore, the catalysts show excellent catalytic activity, high selectivity to higher alcohol, and outstanding stability for the HAS.

Author Contributions: Formal analysis, S.G.; Investigation, S.G., T.H. and Z.Z.; Project administration, G.L. and Y.L.; Resources, Y.L.; Supervision, G.L. and Y.L.; Visualization, S.G.; Writing–original draft, S.G.; Writing–review and editing, G.L.

Funding: This research was funded by National Natural Science Foundation of China [Nos. 21576192 and 21872101] and The APC was funded by Key Science and Technology Program of Henan Province [Nos. 182102210432].

Acknowledgments: The financial support of this work by the National Natural Science Foundation of China (NSFC) (Nos. 21576192 and 21872101), Key Science and Technology Program of Henan Province (Nos. 182102210432) are gratefully acknowledged.

Conflicts of Interest: The authors declare no conflict of interest.

References

1. Baker, J.E.; Burch, R.; Golunski, S.E. Synthesis of higher alcohols over copper/cobalt catalysts. Influence of preparative procedures on the activity and selectivity of Cu/Co/Zn/Al mixed oxide catalysts. *Appl. Catal.* **1989**, *53*, 279–297. [CrossRef]
2. Surisetty, V.R.; Dalai, A.K.; Kozinski, J. Alcohols as alternative fuels: An overview. *Appl. Catal. A-Gen.* **2011**, *404*, 1–11. [CrossRef]
3. Atsumi, S.; Hanai, T.; Liao, J.C. Non-fermentative pathways for synthesis of branched-chain higher alcohols as biofuels. *Nature* **2008**, *451*, 86–90. [CrossRef] [PubMed]
4. Herman, R.G. Advances in catalytic synthesis and utilization of higher alcohols. *Catal. Today* **2000**, *55*, 233–245. [CrossRef]
5. Yu, S.; Tao, J. Economic, energy and environmental evaluations of biomass-based fuel ethanol projects based on life cycle assessment and simulation. *Appl. Energy* **2009**, *86*, S178–S188. [CrossRef]
6. Subramanian, N.D.; Gao, J.; Mo, X.; Goodwin, J.G., Jr.; Torres, W.; Spivey, J.J. La and/or V oxide promoted Rh/SiO_2 catalysts: Effect of temperature, H_2/CO ratio, space velocity, and pressure on ethanol selectivity from syngas. *J. Catal.* **2010**, *272*, 204–209. [CrossRef]

7. Abdelsayed, V.; Shekhawat, D.; Poston, J.A.; Spivey, J.J. Synthesis, characterization, and catalytic activity of Rh-based lanthanum zirconate pyrochlores for higher alcohol synthesis. *Catal. Today* **2013**, *207*, 65–73. [CrossRef]
8. Surisetty, V.; Eswaramoorthi, I.; Dalai, A. Comparative study of higher alcohols synthesis over alumina and activated carbon-supported alkali-modified MoS_2 catalysts promoted with group VIII metals. *Fuel* **2012**, *96*, 77–84. [CrossRef]
9. Christensen, J.M.; Duchstein, L.D.L.; Wagner, J.B.; Jensen, P.A.; Temel, B.; Jensen, A.D. Catalytic Conversion of Syngas into Higher Alcohols over Carbide Catalysts. *Ind. Eng. Chem. Res.* **2012**, *51*, 4161–4172. [CrossRef]
10. Bai, Y.; He, D.; Ge, S.; Liu, H.; Liu, J.; Huang, W. Influences of preparation methods of ZrO_2 support and treatment conditions of Cu/ZrO_2 catalysts on synthesis of methanol via CO hydrogenation. *Catal. Today* **2010**, *149*, 111–116. [CrossRef]
11. Nunan, J.G.; Bogdan, C.E.; Klier, K.; Smith, K.J.; Young, C.W.; Herman, R.G. Methanol and C_2 oxygenate synthesis over cesium doped CuZnO and $Cu/ZnO/Al_2O_3$ catalysts: A study of selectivity and ^{13}C incorporation patterns. *J. Catal.* **1988**, *113*, 410–433. [CrossRef]
12. Fang, K.; Li, D.; Lin, M.; Xiang, M.; Wei, W.; Sun, Y. A short review of heterogeneous catalytic process for mixed alcohols synthesis via syngas. *Catal. Today* **2009**, *147*, 133–138. [CrossRef]
13. Xiao, K.; Bao, Z.; Qi, X.; Wang, X.; Zhong, L.; Fang, K.; Lin, M.; Sun, Y. Structural evolution of CuFe bimetallic nanoparticles for higher alcohol synthesis. *J. Mol. Catal. A-Chem.* **2013**, *378*, 319–325. [CrossRef]
14. Xiao, K.; Qi, X.; Bao, Z.; Wang, X.; Zhong, L.; Fang, K.; Lin, M.; Sun, Y. CuFe, CuCo and CuNi nanoparticles as catalysts for higher alcohol synthesis from syngas: A comparative study. *Catal. Sci. Technol.* **2013**, *3*, 1591–1602. [CrossRef]
15. An, Z.; Ning, X.; He, J. Ga-promoted CO insertion and C–C coupling on Co catalysts for the synthesis of ethanol and higher alcohols from syngas. *J. Catal.* **2017**, *356*, 157–164. [CrossRef]
16. Ning, X.; An, Z.; He, J. Remarkably efficient CoGa catalyst with uniformly dispersed and trapped structure for ethanol and higher alcohol synthesis from syngas. *J. Catal.* **2016**, *340*, 236–247. [CrossRef]
17. Gao, S.; Li, X.; Li, Y.; Yu, H.; Zhang, F.; Sun, Y.; Fang, H.; Zhang, X.; Liang, X.; Yuan, Y. Effects of gallium as an additive on activated carbon-supported cobalt catalysts for the synthesis of higher alcohols from syngas. *Fuel* **2018**, *230*, 194–201. [CrossRef]
18. Zhao, L.; Wei, Y.; Huang, Y.; Liu, Y. $La_{1-x}K_xFe_{0.7}Ni_{0.3}O_3$ catalyst for ethanol steam reforming—The effect of K-doping. *Catal. Today* **2016**, *259*, 430–437. [CrossRef]
19. Morales, M.; Segarra, M. Steam reforming and oxidative steam reforming of ethanol over $La_{0.6}Sr_{0.4}CoO_{3-\delta}$ perovskite as catalyst precursor for hydrogen production. *Appl. Catal. A-Gen.* **2015**, *502*, 305–311. [CrossRef]
20. Lee, M.; Jung, W. Hydrothermal Synthesis of $LaCO_3OH$ and Ln^{3+}-doped $LaCO_3OH$ Powders under Ambient Pressure and Their Transformation to $La_2O_2CO_3$ and La_2O_3. *Bull. Korean Chem. Soc.* **2013**, *34*, 3609–3614. [CrossRef]
21. Kathiraser, Y.; Wang, Z.; Yang, N.-T.; Zahid, S.; Kawi, S. Oxygen permeation and stability study of $La_{0.6}Sr_{0.4}Co_{0.8}Ga_{0.2}O_{3-\delta}$ (LSCG) hollow fiber membrane with exposure to CO_2, CH_4 and He. *J. Membr. Sci.* **2013**, *427*, 240–249. [CrossRef]
22. Levasseur, B.; Kaliaguine, S. Methanol oxidation on $LaBO_3$ (B=Co, Mn, Fe) perovskite-type catalysts prepared by reactive grinding. *Appl. Catal. A-Gen.* **2008**, *343*, 29–38. [CrossRef]
23. Royer, S.; Bérubé, F.; Kaliaguine, S. Effect of the synthesis conditions on the redox and catalytic properties in oxidation reactions of $LaCo_{1-x}Fe_xO_3$. *Appl. Catal. A-Gen.* **2005**, *282*, 273–284. [CrossRef]
24. Fang, Y.; Liu, Y.; Deng, W.; Liu, J. Cu-Co bi-metal catalyst prepared by perovskite $CuO/LaCoO_3$ used for higher alcohol synthesis from syngas. *J. Energy Chem.* **2014**, *23*, 527–534. [CrossRef]
25. Chagas, C.A.; Toniolo, F.S.; Magalhães, R.N.; Schmal, M. Alumina-supported $LaCoO_3$ perovskite for selective CO oxidation (SELOX). *Int. J. Hydrogen Energy* **2012**, *37*, 5022–5031. [CrossRef]
26. Song, Z.; Shi, X.; Ning, H.; Liu, G.; Zhong, H.; Liu, Y. Loading clusters composed of nanoparticles on ZrO_2 support via a perovskite-type oxide of $La_{0.95}Ce_{0.05}Co_{0.7}Cu_{0.3}O_3$ for ethanol synthesis from syngas and its structure variation with reaction time. *Appl. Surf. Sci.* **2017**, *405*, 1–12. [CrossRef]
27. Zhan, H.; Li, F.; Gao, P.; Zhao, N.; Xiao, F.; Wei, W.; Zhong, L.; Sun, Y. Methanol synthesis from CO_2 hydrogenation over La–M–Cu–Zn–O (M = Y, Ce, Mg, Zr) catalysts derived from perovskite-type precursors. *J. Power Sources* **2014**, *251*, 113–121. [CrossRef]

28. Jena, H.; Govindan, K.; Kutty, T. Novel wet chemical synthesis and ionic transport properties of LaGaO$_3$ and selected doped compositions at elevated temperatures. *Mater. Sci. Eng. B* **2004**, *113*, 30–41. [CrossRef]
29. Majima, T.; Kono, E.; Ogo, S.; Sekine, Y. Pre-reduction and K loading effects on noble metal free Co-system catalyst for water gas shift reaction. *Appl. Catal. A-Gen.* **2016**, *523*, 92–96. [CrossRef]
30. Yang, Q.; Cao, A.; Kang, N.; An, K.; Liu, Z.; Liu, Y. A new catalyst of Co/La$_2$O$_3$-doped La$_4$Ga$_2$O$_9$ for direct ethanol synthesis from syngas. *Fuel Process. Technol.* **2018**, *179*, 42–52. [CrossRef]
31. Boz, I. Higher alcohol synthesis over a K-promoted Co$_2$O$_3$-CuO-ZnO-Al$_2$O$_3$. *Catal. Lett.* **2003**, *87*, 187–194. [CrossRef]
32. Courty, P.; Durand, D.; Freund, E.; Sugeer, A. C1-C6 alcohols from synthesis gas on copper-cobalt catalysts. *J. Mol. Catal.* **1982**, *17*, 241–254. [CrossRef]
33. Sheffer, G.R.; Jacobson, R.A.; King, T.S. Chemical nature of alkali-promoted copper-cobalt-chromium oxide higher alcohol catalysts. *J. Catal.* **1989**, *116*, 95–107. [CrossRef]
34. Sheffer, G.R.; King, T.S. Effect of preparation parameters on the catalytic nature of potassium promoted CuCoCr higher alcohol catalysts. *Appl. Catal.* **1988**, *44*, 153–164. [CrossRef]
35. Wang, J.; Chernavskii, P.A.; Khodakov, A.Y.; Wang, Y. Structure and catalytic performance of alumina-supported copper–cobalt catalysts for carbon monoxide hydrogenation. *J. Catal.* **2012**, *286*, 51–61. [CrossRef]
36. Liu, G.; Niu, T.; Pan, D.; Liu, F.; Liu, Y. Preparation of bimetal Cu–Co nanoparticles supported on meso–macroporous SiO$_2$ and their application to higher alcohols synthesis from syngas. *Appl. Catal. A-Gen.* **2014**, *483*, 10–18. [CrossRef]
37. Tien-Thao, N.; Zahedi-Niaki, M.H.; Alamdari, H.; Kaliaguine, S. Conversion of syngas to higher alcohols over nanosized LaCo$_{0.7}$Cu$_{0.3}$O$_3$ perovskite precursors. *Appl. Catal. A-Gen.* **2007**, *326*, 152–163. [CrossRef]
38. Dong, X.; Liang, X.; Li, H.; Lin, G.; Zhang, P.; Zhan, H. Preparation and characterization of carbon nanotube-promoted Co–Cu catalyst for higher alcohol synthesis from syngas. *Catal. Today* **2009**, *147*, 158–165. [CrossRef]
39. Liu, G.; Cui, J.; Luo, R.; Liu, Y.; Huang, X.; Wu, N.; Jin, X.; Chen, H.; Tang, S.; Kim, J.; et al. 2D MoS$_2$ grown on biomass-based hollow carbon fibers for energy storage. *Appl. Surf. Sci.* **2019**, *469*, 854–863. [CrossRef]
40. Tsipouriari, V.A.; Verykios, X.E. Carbon and Oxygen Reaction Pathways of CO$_2$ Reforming of Methane over Ni-La$_2$O$_3$ and Ni-Al$_2$O$_3$ Catalysts Studied by Isotopic Tracing Techniques. *J. Catal.* **1999**, *187*, 85–94. [CrossRef]
41. Verykios, X.E. Catalytic dry reforming of natural gas for the production of chemicals and hydrogen. *Int. J. Hydrogen Energy* **2003**, *106*, 1045–1063. [CrossRef]
42. Li, S.; Tang, H.; Gong, D.; Ma, Z.; Liu, Y. Loading Ni/La$_2$O$_3$ on SiO$_2$ for CO methanation from syngas. *Catal. Today* **2017**, *297*, 298–307. [CrossRef]
43. Liu, G.; Geng, Y.; Pan, D.; Zhang, Y.; Niu, T.; Liu, Y. Bi-metal Cu–Co from LaCo$_{1-x}$Cu$_x$O$_3$ perovskite supported on zirconia for the synthesis of higher alcohols. *Fuel Process. Technol.* **2014**, *128*, 289–296. [CrossRef]
44. Liu, J.; Li, C.; Wang, F.; He, S.; Chen, H.; Zhao, Y.; Wei, M.; Evans, D.G.; Duan, X. Enhanced low-temperature activity of CO$_2$ methanation over highly-dispersed Ni/TiO$_2$ catalyst. *Catal. Sci. Technol.* **2013**, *3*, 2627–2633. [CrossRef]
45. Guo, J.; Hou, Z.; Gao, J.; Zheng, X. Production of Syngas via Partial Oxidation and CO$_2$ Reforming of Coke Oven Gas over a Ni Catalyst. *Energy Fuels* **2008**, *22*, 1444–1448. [CrossRef]

© 2019 by the authors. Licensee MDPI, Basel, Switzerland. This article is an open access article distributed under the terms and conditions of the Creative Commons Attribution (CC BY) license (http://creativecommons.org/licenses/by/4.0/).

Article

Segregation of Nickel/Iron Bimetallic Particles from Lanthanum Doped Strontium Titanates to Improve Sulfur Stability of Solid Oxide Fuel Cell Anodes

Patrick Steiger [1,2], Dariusz Burnat [3], Oliver Kröcher [1,2], Andre Heel [3] and Davide Ferri [1,*]

[1] Paul Scherrer Institut, Forschungsstrasse 111, CH-5232 Villigen, Switzerland; patrick.steiger@psi.ch (P.S.); oliver.kroecher@psi.ch (O.K.)
[2] École Polytechnique Fédérale de Lausanne (EPFL), Institute of Chemical Sciences and Engineering, CH-1015 Lausanne, Switzerland
[3] Zurich University of Applied Sciences, IMPE – Institute for Materials and Process Engineering, CH-8400 Winterthur, Switzerland; dariuszartur.burnat@zhaw.ch (D.B.); andre.heel@zhaw.ch (A.H.)
* Correspondance: davide.ferri@psi.ch; Tel.: +41-56-310-2781

Received: 19 March 2019; Accepted: 1 April 2019; Published: 3 April 2019

Abstract: Perovskite derived Ni catalysts offer the remarkable benefit of regeneration after catalyst poisoning or Ni particle growth through the reversible segregation of Ni from the perovskite-type oxide host. Although this property allows for repeated catalyst regeneration, improving Ni catalyst stability towards sulfur poisoning by H_2S is highly critical in solid oxide fuel cells. In this work Mn, Mo, Cr and Fe were combined with Ni at the B-site of $La_{0.3}Sr_{0.55}TiO_{3\pm\delta}$ to explore possible benefits of segregation of two transition metals towards sulfur tolerance. Catalytic activity tests towards the water gas shift reaction were carried out to evaluate the effect of the additional metal on the catalytic activity and sulfur stability of the Ni catalyst. The addition of Fe to the Ni perovskite catalyst was found to increase sulfur tolerance. The simultaneous segregation of Fe and Ni from $La_{0.3}Sr_{0.55}Ti_{0.95-x}Ni_{0.05}Fe_xO_{3\pm\delta}$ ($x \leq 0.05$) was investigated by temperature programmed reduction, X-ray diffraction and X-ray absorption spectroscopy and catalytic tests after multiple redox cycles. It is shown that catalytic properties of the active phase were affected likely by the segregation of Ni/Fe alloy particles and that the reversible segregation of Ni persisted, while it was limited in the case of Fe under the same conditions.

Keywords: nickel; $La_{0.3}Sr_{0.55}Ti_{0.95}Ni_{0.05}O_{3\pm\delta}$; catalyst regeneration; structural reversibility; H_2S; solid oxide fuel cell

1. Introduction

In recent years perovskite-type metal oxide (PMO) derived metal catalysts have attracted great attention for their high redox stability, due to the reversible segregation of catalytically active metals from the bulk of the oxide in reducing atmospheres and their reincorporation during oxidative treatments [1]. It was demonstrated that this property allows for the regeneration of catalysts, which have suffered from active metal particle sintering, as well as the recovery of catalysts poisoned by coke or sulfur through simple redox cycling [2,3]. However, achieving catalyst stability while maintaining high catalytic conversion rates, thus decreasing the necessary frequency of catalyst regeneration cycles, appears to be as propitious as increasing catalyst regenerability. This is especially important in redox-sensitive electrochemical devices, such as solid oxide fuel cells (SOFCs), where metallic Ni is typically applied as the active phase in the anode for fuel oxidation, but also for its activity towards the water gas shift reaction (WGS) when the device is operated on CO-rich feeds [4,5]. Prominent examples of sulfur poisoning in heterogeneous catalysis include exhaust gas after-treatment reactions in the three-way catalytic converters and the selective catalytic reduction of NO_x compounds [6], methanation of carbon oxides [7], reforming of methane and higher hydrocarbons [8,9], Fischer-Tropsch [10] and methanol synthesis from syngas [11]. In the case of all metal-catalyzed reactions, sulfur tolerance may generally be increased in three ways: (i) Increasing the number of catalytically active sites, which leaves higher number of free active sites at equal sulfur surface coverage, (ii) a sacrificial species may be introduced on the catalyst, which preferentially interacts with sulfur leaving the active species available for the reaction and (iii) the electronic effect on the active metal caused by the introduction of a second metal may result in decreased metal-sulfur interactions [12]. In the case of the Ni-catalyzed WGS reaction, promising results have been reported regarding improved sulfur stability of Ni reforming catalysts by the addition of Mo, Co and Re [13]. The beneficial effect of Re was attributed to the formation of a sulfur tolerant alloy, whereas in Ni-Mo metal combinations Mo acted as the sacrificial element [14]. The interaction between Ni and Mo was found to also increase the electron density on Mo thus facilitating its interaction strength with electronegative sulfur. Re-doping was also applied to improve sulfur tolerance of a Ni-Sr/ZrO_2 catalyst for the reforming of hydrocarbons [15]. Metal-metal interactions were also exploited to reduce the electron donor capacity of Pd by Mn addition thus decreasing its interaction strength with sulfur [16–18]. It is likely that doping of the Ni phase with other transition metals may also change sulfur adsorption properties on Ni.

It is the aim of the present work to combine the excellent regenerability of PMO-derived Ni catalysts for the WGS reaction with the possibility to alter the adsorption properties by transition metal doping of $La_{0.3}Sr_{0.55}Ti_{0.95}Ni_{0.05}O_{3\pm\delta}$, a self-regenerable SOFC anode material [2]. Sulfur sensitive elements, such as Cr, Mn, Fe and Mo were selected as potential sacrificial agents for screening towards a Ni-metal combination resistant to sulfur. Molybdenum, Cr and Mn are also of great importance to industrial high-temperature WGS catalysts [19].

2. Results

The work is structured as follows. We start showing the characterization of all materials (Table S1 for sample denotation) using ex situ and in situ X-ray diffraction (XRD) and temperature programmed reduction (TPR). Then the catalytic activity of the materials towards the water gas shift reaction (WGS) in the absence and presence of H_2S is presented. Because $La_{0.3}Sr_{0.55}Ti_{0.95}Ni_{0.05}O_{3\pm\delta}$ impregnated by Fe resulted in the most promising in terms of sulfur uptake and resistance to poisoning, further samples were prepared with various Fe/Ni ratios (see Materials and Methods and Table 1 for sample denotation), which were characterized also for the local environment of Ni and Fe using X-ray absorption spectroscopy and were tested for reaction and poisoning.

2.1. Characterization

Figure 1 shows X-ray diffraction (XRD) patterns of the materials after impregnation of $La_{0.3}Sr_{0.55}TiO_{3\pm\delta}$ (LST) and $La_{0.3}Sr_{0.55}Ti_{0.95}Ni_{0.05}O_{3\pm\delta}$ (LSTN) with the metal precursors and subsequent calcination. XRD patterns are also given for LST (Figure 1a) and LSTN (Figure 1b). The reflections of the corresponding single metal oxides were observed in the XRD patterns of impregnated LST (LST-5Me, Me = Cr, Mn, Fe and Ni; Figure 1c), as well as impregnated LSTN (LSTN-5Me, Me = Cr, Mn and Fe; Figure 1d). Impregnation with the Mo precursor (LST-5Mo and LSTN-5Mo) resulted in the presence of reflections of a $SrMoO_4$ phase, which is indicative of the segregation of small quantities of Sr from both LST and LSTN lattices during impregnation and calcination.

TPR experiments revealed the reduction of MnO, Cr_2O_3, Fe_2O_3 and NiO on LST up to 800 °C (Figure 2a). $SrMoO_4$ was only partially reduced at these temperatures as is evident from the fact that H_2 was still consumed at 800 °C. Similar behavior was also observed on impregnated LSTN materials (Figure 2c), on which also contributions of the LSTN support can be observed between ca. 450 °C and 650 °C. XRD measurements of the reduced materials confirmed the reduction of NiO, Fe_2O_3 and partial reduction of $SrMoO_4$ on LST-5Ni, LST-5Fe and LST-5Mo, respectively (Figure 2b). No reflections of Cr and Mn metals were observed on reduced LST-5Cr and LST-5Mn, respectively. Instead, reflections of the single oxides were observed, similar to the calcined materials. This was likely due to the ex situ nature of the XRD experiments and the strong tendency of dispersed Cr and Mn particles to form oxides.

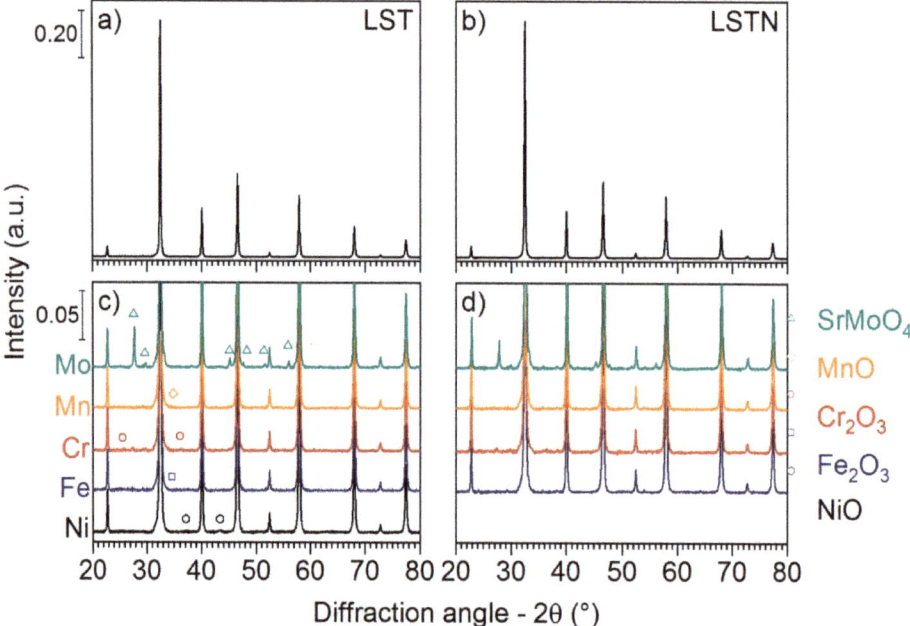

Figure 1. Powder XRD patterns of (**a**) La$_{0.3}$Sr$_{0.55}$TiO$_{3\pm\delta}$ (LST), (**b**) La$_{0.3}$Sr$_{0.55}$Ti$_{0.95}$Ni$_{0.05}$O$_{3\pm\delta}$ (LSTN), (**c**) LST-5Me (Me = Mo, Mn, Cr, Fe, Ni) and (**d**) LSTN-5Me (Me = Mo, Mn, Cr and Fe). Markers indicate the presence of metal oxide phases after impregnation. Note that the intensity of the diffractograms in (**c**) and (**d**) is magnified (5×) with respect to (**a**) and (**b**).

All reflections were also encountered on the reduced LSTN-type materials. However, reduced LSTN-5Fe exhibited reflections that indicated the presence of two Fe allotropes (α-Fe and γ-Fe, Figure 2d), which is likely a consequence of Fe/Ni alloy formation during reduction at 800 °C as it was not observed on LST-5Fe. Subsequent rapid cooling to room temperature after reduction resulted in phase separation. This is supported by phase diagrams of the Fe-Ni system, which predict partial phase decomposition of the homogeneous γ-Fe/Ni alloy phase and various phase transformations during cooling [20]. Since cooling rates were high in these experiments (ca. 20 °C·min^{-1}) the presence of metallic phases, which are not at equilibrium is highly probable [21]. Ni (111) reflections could be observed in LSTN, LSTN-5Cr and LSTN-5Mn, whereas the absence of the same in LSTN-5Fe and LSTN-5Mo can be regarded as an indication of Ni/Me alloy formation in the latter cases.

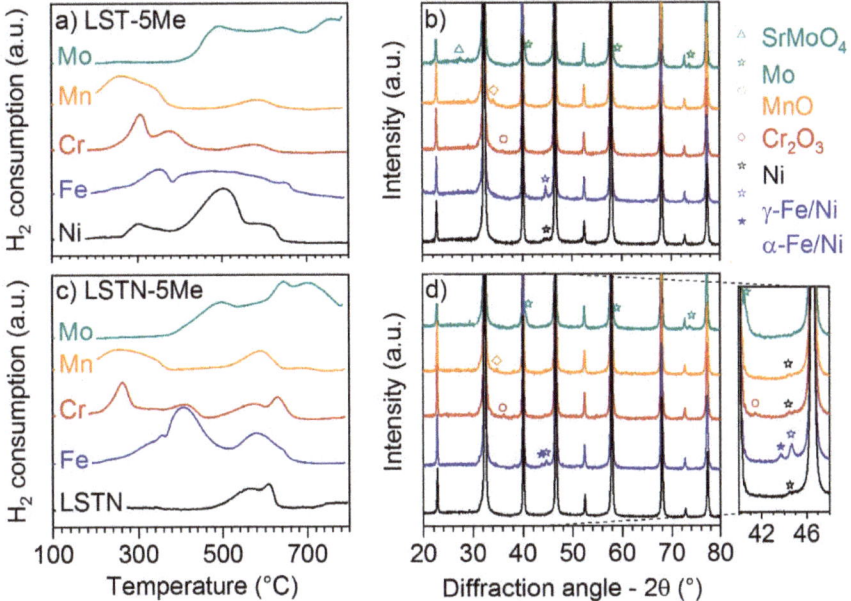

Figure 2. TPR profiles of (**a**) LST-5Me (Me = Mo, Mn, Cr, Fe, Ni) and (**c**) LSTN-5Me (Me = Mo, Mn, Cr, Fe). (**b**) Powder XRD patterns of reduced LST-type materials and (**d**) LSTN-type materials (10 vol.% H_2/Ar, 800 °C, 1 h). The additional panel in (**d**) displays the magnified angular range $40° \leq 2\theta \leq 48°$ (3×) with respect to intensity.

2.2. Catalytic Activity

The activity of the pre-reduced catalysts towards the water gas shift (WGS) reaction was assessed by measuring CO conversion between 300 °C and 800 °C and the results are shown in Figure 3. Activities of the catalysts produced using Ni-free LST as a support are displayed in Figure 3a. Nickel was the most active among all metals and LST-5Ni exhibited CO conversion above 60% at 460 °C. The next best activities were shown by LST-5Fe followed by LST-5Mn and finally LST-5Mo and LST-5Cr, which showed only limited activities at reactions temperatures below 800 °C. The lack of activity of Cr and Mo is not surprising. Molybdenum was added with the specific intent to introduce an element with negligible WGS activity, but with significant sensitivity to sulfur to be used as a sacrificial agent. Chromium is mainly used in Fe/Cr-based WGS catalysts for its capability to stabilize the active Fe phase [22]. Since LST did not show any significant WGS activity (Figure 3a) it can be assumed that the catalytic activity of LST-Ni, LST-Fe and LST-Mn was due to the impregnated metal. The CO conversion of each catalyst after poisoning is shown by the dashed lines in Figure 3a. Only LST-5Ni suffered severely during catalyst poisoning using 50 ppm H_2S under reaction conditions (800 °C, 1 h). The catalytic activity of LST-5Fe, LST-5Mn, LST-5Cr and LST-5Mo on the other hand, remained rather constant.

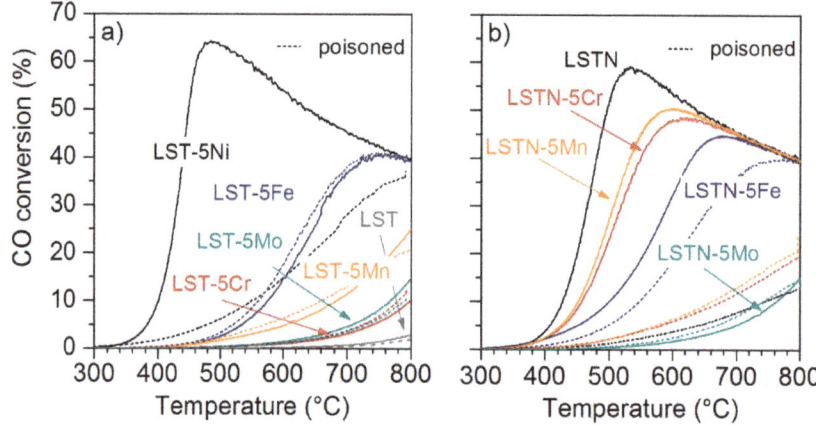

Figure 3. (**a**) WGS activity of LST impregnated with Cr (red), Fe (blue), Mn (orange), Mo (turquoise) and Ni (black) before (continuous lines) and after (dotted lines) catalyst poisoning with 50 ppm H_2S. (**b**) WGS activity of LSTN (black), as well as LSTN impregnated with Cr (red), Fe (blue), Mn (orange) and Mo (turquoise).

Catalytic activities of impregnated LSTN-type catalysts are displayed in Figure 3b. LSTN exhibited the highest activity towards the WGS reaction, followed by LSTN-5Mn and LSTN-5Cr. LSTN-5Fe also exhibited higher activity than its Ni-free counterpart LST-5Fe, whereas the low activity of LSTN-5Mo remained unchanged by the presence of Ni in the sample. The catalytic results provide an indication for Ni alloying with the impregnated metals Cr, Fe and Mn. The fact that all LSTN-5Me (Me = Cr, Fe, Mn, Mo) catalysts exhibited lower catalytic activity than LSTN, while all metals showed at least some activity towards WGS on Ni-free LST, provides evidence for close interaction between segregated Ni and Me in LSTN-5Me or coverage of the active Ni phase with the less active Me. This interaction was found to be less beneficial for WGS activity under sulfur-free conditions. However, catalytic tests after sulfur poisoning showed a stabilizing effect of the metals on WGS activity. Especially LSTN-5Fe was able to maintain comparably high levels of catalytic activity after poisoning by H_2S. It should be noted that the activity of LSTN-5Fe after poisoning did not exceed the one of LST-5Fe, which could indicate that Ni is still poisoned. Nevertheless, the improved activity of LSTN-5Fe in sulfur-free conditions compared to LST-5Fe demonstrates a potential benefit of a bimetallic catalyst.

2.3. Sulfur Uptake

Differences between the various materials were also observed in their behavior during H_2S adsorption. Figure 4a shows H_2S breakthrough curves for LST-5Me catalysts during sulfur loading. H_2S (50 ppm) was introduced to the reaction gas stream after 5 min equilibration time. Dosing H_2S over a blank quartz reactor resulted in negligible retention time and 50 ppm were attained after ca. 5 min. Sulfur adsorption can be observed by the increased retention times when H_2S is dosed over the catalyst bed. The longest retention time of around 8 min was recorded for LST-5Ni. The significant sulfur uptake of this sample (210 ppm by weight) is in line with its strong deactivation in terms of the catalytic activity after sulfur loading (Figure 3a). A retention time of around 4 min and 2 min was observed for LST-5Fe and LST-5Mn, respectively. H_2S breakthrough of LST-5Mo and LST-5Cr was close to the blank experiment indicating low sulfur adsorption properties of these metals. This indicates that the ability of Mo as a sulfur scavenger was limited on these samples compared to the more conventionally applied single oxides MoO_2 and MoO_3, possibly due to the $SrMoO_4$ phase formed during synthesis (Figure 1).

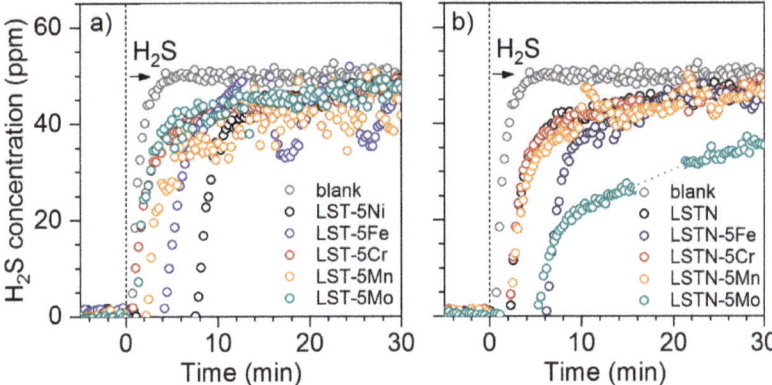

Figure 4. H$_2$S breakthrough curves during H$_2$S adsorption experiments on reduced (**a**) LST-5Me (Me = Mo, Mn, Cr, Fe, Ni) and (**b**) LSTN-5Me (Me = Mo, Mn, Cr and Fe). The start time of sulfur addition is indicated by the vertical dashed line.

Interestingly, LSTN-5Mo showed significant sulfur uptake exceeding the combined sulfur storage capabilities of LSTN and LST-5Mo. This can be regarded as an indication for close interaction between segregated Ni and Mo, which appeared to significantly change the sulfur uptake properties of the material. Retention times of LSTN-5Fe, LSTN-5Cr and LSTN-5Mo were approximately the sum of those of LSTN and their LST-5Me counterparts and can be therefore explained by the increased metal content on the sample surface. Sulfur uptake of LSTN was much lower than that of LST-5Ni (ca. 40 ppm compared to 210 ppm), which can be justified using the similar argumentation that when Ni is deposited on LST by impregnation, the Ni metal surface is higher compared to pre-reduced LSTN, due to partial Ni reduction in the latter. Although no obvious advantage of metal impregnation on LSTN could be observed from these H$_2$S adsorption experiments because none of the metals reduced the H$_2$S uptake of the sample, the increased sulfur tolerance in terms of the catalytic activity towards WGS exhibited by LSTN-5Fe (Figure 3b) could be an advantageous property. Therefore, the question arises whether it is possible to exploit the reversible segregation of metals from an LST-type host to produce both redox stable, as well as sulfur tolerant Ni/Fe catalysts.

2.4. $La_{0.3}Sr_{0.55}Ti_{1-x-y}Fe_xNi_yO_{3\pm\delta}$ (x = 0, 0.025, 0.05; y = 0, 0.05)

Catalytic activity and sulfur uptake data indicated that Fe might be a suitable candidate to improve sulfur tolerance of LSTN while maintaining the self-regeneration property. In order to explore the potential segregation of both Fe and Ni from the perovskite-type host, both Ni and Fe were introduced at the perovskite B-site of LST to obtain compositions of the type $La_{0.3}Sr_{0.55}Ti_{1-x-y}Fe_xNi_yO_{3\pm\delta}$ (x = 0, 0.025, 0.05; y = 0, 0.05) the denotations of which are summarized in Table 1. Figure 5 displays a summary of XRD patterns collected on calcined materials, as well as after 15 h reduction at 800 °C (10 vol.% H$_2$/Ar).

All calcined samples exhibited only reflections, which could be attributed to the perovskite host. After reduction, weak reflections belonging to metallic phases were observed. Reduced $La_{0.3}Sr_{0.55}Ti_{0.95}Ni_{0.05}O_{3\pm\delta}$ (LSTN-5Ni) showed a reflection centered at 44.40°, which corresponds to the Ni (111) reflection. Reduced $La_{0.3}Sr_{0.55}Ti_{0.925}Fe_{0.025}Ni_{0.05}O_{3\pm\delta}$ (LSTFN-2Fe5Ni) and $La_{0.3}Sr_{0.55}Ti_{0.9}Fe_{0.05}Ni_{0.05}O_{3\pm\delta}$ (LSTFN-2Fe5Ni) exhibited a weak reflection at 43.77°, which is higher than one would expect for the (111) reflection of metallic Fe (43.6° 2θ), but certainly at lower angles than the Ni (111) reflection. This can be regarded as evidence for Ni/Fe alloy particle formation upon reduction of LSTFN-type materials. $La_{0.3}Sr_{0.55}Ti_{0.925}Fe_{0.025}Ni_{0.05}O_{3\pm\delta}$ (LSTF-5Fe) did not show significant reflections of metallic Fe after reduction. However, a new reflection appeared at 38.27°, which could not be assigned to any metallic phase. While metallic Ti is expected to display a reflection at around 38.4°, it is highly unlikely that it formed under these pretreatment conditions, due to the inherent stability of Ti^{4+}. The missing reflection at 35.2° excludes this possibility conclusively. However, the observed reflection could be explained by the formation of a new perovskite-type phase of lower symmetry. Orthorhombic perovskites (such as A-site stoichiometric $LaSrFeO_{3\pm\delta}$) possess a reflection at ca. 38° corresponding to the (113) lattice planes. All other reflections might be either hidden below the dominant original perovskite phase or too weak to be observed.

Table 1. Sample denotations and metal content of $La_{0.3}Sr_{0.55}Ti_{1-x-y}Fe_xNi_yO_{3\pm\delta}$ -type samples.

Denotation	Formula	Ni Content (mol.%)	Fe Content (mol.%)
LSTN-5Ni	$La_{0.3}Sr_{0.55}Ti_{0.95}Ni_{0.05}O_{3\pm\delta}$	5	0
LSTFN-2Fe5Ni	$La_{0.3}Sr_{0.55}Ti_{0.925}Fe_{0.025}Ni_{0.05}O_{3\pm\delta}$	5	2.5
LSTFN-5Fe5Ni	$La_{0.3}Sr_{0.55}Ti_{0.9}Fe_{0.05}Ni_{0.05}O_{3\pm\delta}$	5	5
LSTF-5Fe	$La_{0.3}Sr_{0.55}Ti_{0.95}Fe_{0.05}O_{3\pm\delta}$	0	5

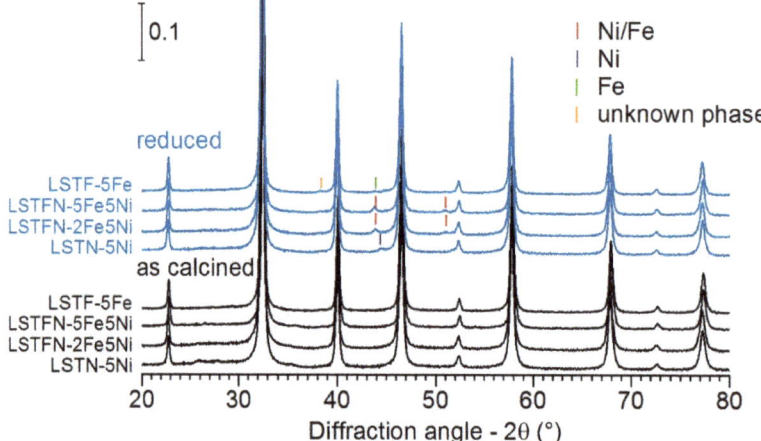

Figure 5. Powder XRD patterns of calcined and reduced LSTN-5Ni, LSTFN-2Fe5Ni, LSTFN-5Fe5Ni and LSTF-5Fe (10 vol.% H_2/Ar, 800 °C, 15 h). Diffractograms are magnified with respect to intensity (2×) to emphasize Ni, Fe and Ni/Fe reflections.

The materials were subjected to TPR and reoxidation cycles to verify the structural reversibility and to determine the temperature at which segregated metals are reversibly incorporated into the host perovskite lattice. In these experiments, TPR profiles were followed by an isothermal reduction at 800 °C. The materials were then subsequently reoxidised at the indicated temperature (700 °C, 750 °C, 800 °C, 850 °C, 900 °C and 950 °C) before the next TPR profile was collected. The TPR profile is sensitive to the nature and coordination environment of reducible metal species and such experiments can, therefore, be exploited to determine the reoxidation temperature needed to reestablish the state of the reducible metal species in the initial calcined material [2,23]. Figure 6a displays the TPR redox cycles obtained on LSTN-5Ni. The reduction feature of NiO (ca. 370 °C) disappears after reoxidation at $T_{reox} \geq 800$ °C, thus indicating successful and complete Ni reincorporation at this temperature [2]. The initial TPR of calcined LSTFN-2Fe5Ni (Figure 6b) was not as well defined as the one recorded for LSTN-5Ni. Instead of the distinct double feature, the sample exhibited a broad reduction peak between 400 °C and 650 °C. After reduction and subsequent reoxidation at low temperatures (700 °C and 750 °C) the sample exhibited a low temperature feature peaking at around 475 °C. This feature then transitioned into the previously observed double feature for reoxidation above 850 °C, which could be interpreted as the temperature at which both Fe and Ni are reincorporated into the perovskite host.

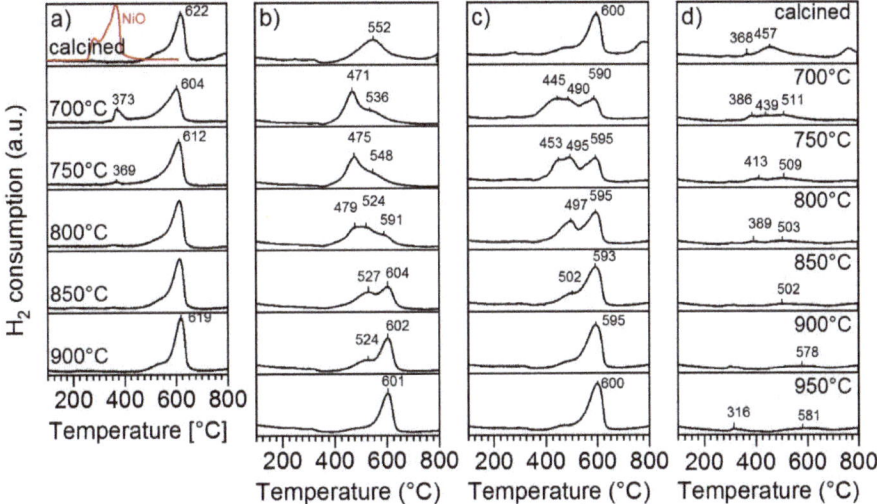

Figure 6. TPR reduction-reoxidation cycles for (**a**) LSTN, (**b**) LSTFN-2Fe5Ni, (**c**) LSTFN-5Fe5Ni and (**d**) LSTF-5Fe. Hydrogen consumption values were normalized by molar quantity of sample. Reoxidized samples were subjected to pre-reduction (10 vol.% H_2/Ar, 800 °C, 1 h) before reoxidation (20 vol.% O_2/N_2, 2 h) at the temperature indicated for each row.

The sample with higher Fe content (LSTFN-5Fe5Ni; Figure 6c) exhibited the reduction feature, which was attributed previously to a two-step reduction process of Ni [24]. However, peak reduction temperatures were shifted to lower temperatures by 22 °C compared to the ones recorded for LSTN-5Ni in Figure 6a. Reoxidation at lower temperatures caused the formation of a new reduction feature between 350 °C and 550 °C also on this sample, which can be attributed to the reoxidation of Ni/Fe oxides at the perovskite surface. Reestablishment of the initial reduction profile was achieved after reoxidation at 850 °C. Interestingly, LSTF-5Fe (Figure 6d) did not exhibit as an extensive reduction as the other samples; the reduction features were broad and attenuated with increasing reoxidation temperature. In this case, TPR-redox cycling seemed unsuitable to accurately trace the reoxidation temperature necessary for Fe reincorporation and demonstrates the limited reducibility of Fe in LSTF-5Fe in absence of Ni.

Even though TPR provides important insight in the reducibility of the materials, only an element specific method, such as X-ray absorption spectroscopy (XAS) can be used to ultimately differentiate between individual contributions of two or more reducible species. Therefore, XAS was applied to investigate the effect of the simultaneous presence of both Fe and Ni in LSTFN-type samples on the reduction and reoxidation of the individual metals. Figure 7 displays the Ni K-edge (8.333 keV) X-ray absorption near edge structure (XANES) spectra of the Ni-containing samples LSTN-5Ni (Figure 7a), LSTFN-2Fe5Ni (Figure 7b) and LSTFN-5Fe5Ni (Figure 7c) in their calcined state, as well as after reduction (10 vol.% H_2/Ar, 800 °C, 15 h) and reoxidation (20 vol.% O_2/N_2, 800 °C, 2 h).

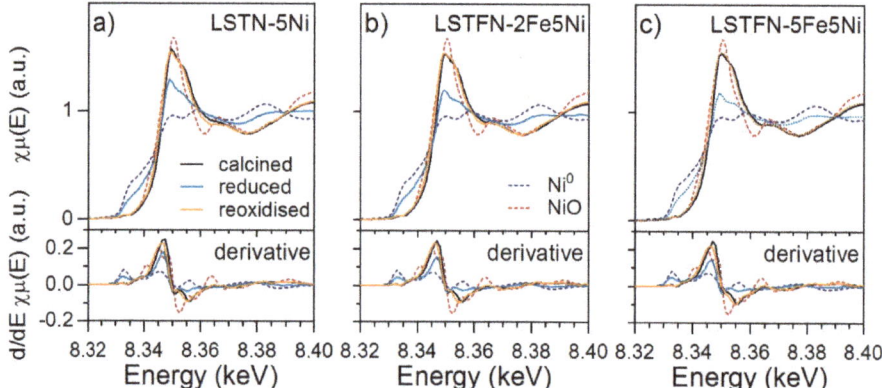

Figure 7. Normalized Ni K-edge (8.333 keV) XANES spectra of (**a**) calcined and reduced (10 vol.% H_2/Ar, 800 °C, 15 h) LSTN-5Ni, (**b**) LSTFN-2Fe5Ni and (**c**) LSTFN-5Fe5Ni. Plots of the first derivative of the normalized spectra are shown in the bottom panels. The spectra of Ni foil and NiO reference materials are included for comparison (dashed lines).

The Ni K-edge spectra of the calcined materials displayed an intense whiteline with a high energy shoulder that can be taken as characteristic of Ni adopting the coordination environment of Ti [2,25,26]. The edge energy was also higher than in the case of Ni^{2+} in NiO. After reduction, the spectra corresponded to a linear combination of the spectra of Ni adopting the coordination environment of the B-site after calcination and Ni^0. Although TPR analysis showed reversibility only after reoxidation at 850 °C (Figure 6), the shape of the XANES spectra and thus the state of Ni were completely reversible in this redox cycle, which was carried out at lower temperatures. It was also observed that the contribution of the Ni^0 reference to the spectra of the reduced samples increased with increasing Fe concentration. This was confirmed by linear combination fit (LCF) of the spectra indicating that the amount of Ni^0 increased from 52% in LSTN-5Ni to 62% in LSTFN-2Fe5Ni and to 67% in LSTFN-5Fe5Ni. LCF results are summarized in Figure S1 and the corresponding fit results are shown in Figure S2.

Evidence that Ni was not only reduced, but also segregated and formed metallic particles is provided by the extended X-ray absorption fine structure of the Ni K-edge. The k^3-weighted data is shown in Figure S3 for all samples, as well as Ni references. The radial distances of coordination shells become obvious through Fourier transformation of this data, as shown in Figure 8. After reduction, the feature attributed to a Ni-Ni coordination shell appeared at 2.15 Å. This feature was present for all samples so that metal particle formation can be assumed likewise.

Fe K-edge (7.112 keV) XANES data was obtained on the Fe-containing samples LSTFN-2Fe5Ni, LSTFN-5Fe5Ni and LSTF-5Fe and spectra of calcined, reduced and reoxidized samples are displayed in Figure 9 along with spectra of Fe^0, FeO, Fe_3O_4 and Fe_2O_3 reference compounds. Significant changes in the shape of the XANES could be observed also for Fe K-edge absorption spectra over the redox cycle. Clear Fe reduction could be observed by a decrease in whiteline intensity, as well as a shift in absorption edge energy (E_0). This shift is best determined through the position of the first maximum in the derivative of the absorption curves, which changes from 7.128 keV for calcined materials to 7.125 keV for reduced materials, corresponding to a decreased ionization energy, which is typically observed for reduced states. However, in the spectra of the reduced samples, the whiteline did not correspond to a simple linear combination of the reference spectrum of Fe^0 and the spectrum of calcined LSTFN- or LSTF-type materials, thus suggesting other states of Fe.

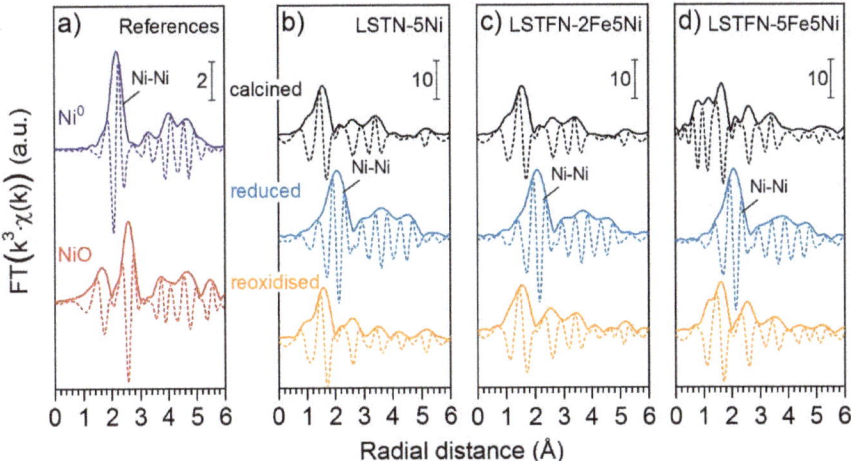

Figure 8. Fourier transformed k³-weighted Ni K-edge EXAFS data obtained for (**a**) Ni⁰ (Ni foil) and NiO reference materials, as well as calcined, reduced (10 vol.% H_2/Ar, 800 °C, 1 h) and reoxidized (20 vol.% O_2/N_2, 800 °C, 2 h) (**b**) LSTN-5Ni, (**c**) LSTFN-2Fe5Ni and (**d**) LSTFN-5Fe5Ni. Features are labelled on the reduced materials according to the underlying Ni-Ni scattering paths.

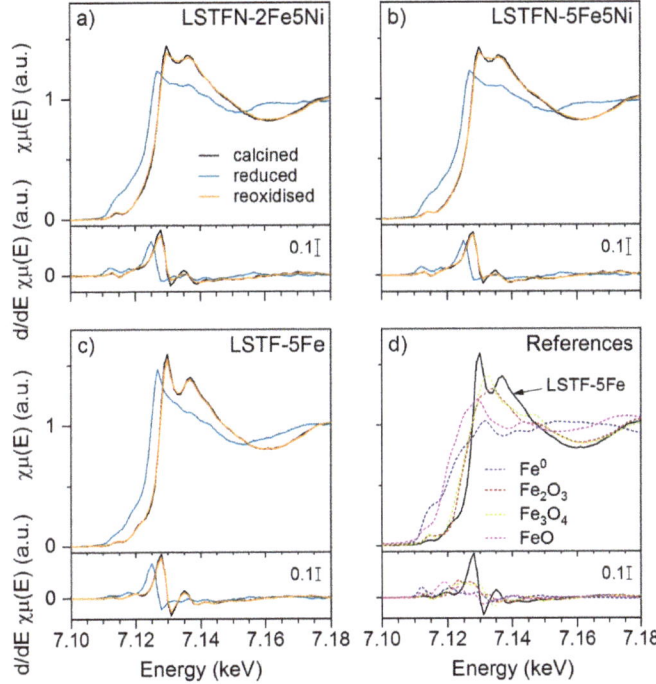

Figure 9. Normalized Fe K-edge (7.112 keV) XANES spectra of (**a**) calcined, reduced (800 °C, 10 vol.% H_2/Ar, 15 h) and reoxidized (800 °C, 20 vol.% O_2/N_2, 2 h) LSTFN-2Fe5Ni, (**b**) LSTFN-5Fe5Ni, (**c**) LSTF-5Fe. (**d**) XANES spectra of LST-5Fe, Fe⁰, Fe_2O_3, Fe_3O_4 and FeO reference materials. The first derivative of the normalized spectra is shown in the bottom panels.

No suitable fits could be obtained through LCF analysis using all displayed Fe reference spectra. This may be linked to the presence of other Fe-containing phases as was already suggested from the XRD patterns in Figure 1. Furthermore, the XANES of LSTFN-2Fe5Ni and LSTFN-5Fe5Ni was not completely restored in the reoxidized materials as can be seen in the region of the local minimum at around 7.135 keV. LCF of the spectra of the reoxidized materials indicated the presence of Fe_3O_4 (ca. 17%) and thus incomplete reincorporation of Fe under the applied reoxidation conditions.

The k^3-weighted Fe K-edge (7.112 keV) EXAFS data of LSTFN-2Fe5Ni, LSTFN-5Fe5Ni and LSTF-5Fe is shown in Figure S4, whereas the Fourier transformed data is shown in Figure 10. Compared to the Ni^0 reference in Figure 8, Fe^0 in the Fe foil displayed the first coordination shell at a slightly longer radial distance (2.23 Å) and similar to the Ni K-edge data contributions of this feature, could be found in the spectra of the reduced samples. This indicates that besides Ni Fe was also partially reduced to Fe^0 and was present in the form of metal particles. Interestingly, the contribution of this feature to the spectra of the reduced samples decreased with decreasing Ni/Fe ratios along the series LSTFN-2Fe5Ni > LSTFN-5Fe5Ni > LSTF-5Fe, which suggests that larger Ni content favors Fe reduction. Hence, the positive influence of one metal on the reducibility and segregation of the other metal could be observed.

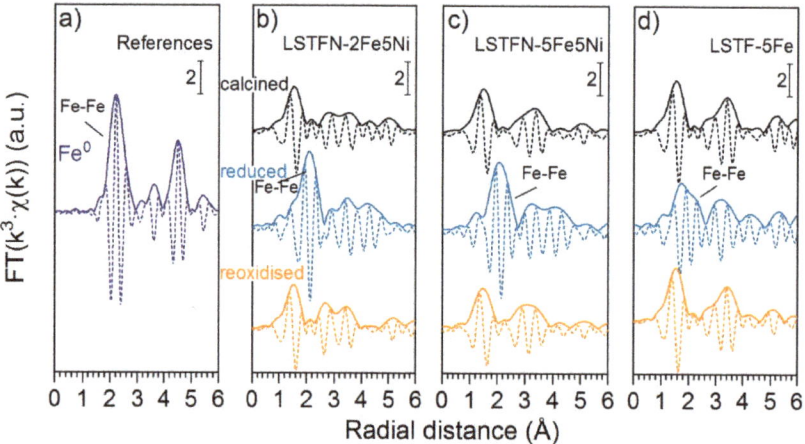

Figure 10. Fourier transformed k^3-weighted Fe K-edge EXAFS data obtained for (**a**) Fe^0 (Fe foil) reference, as well as (**b**) LSTFN-2Fe5Ni, (**c**) LSTFN-5Fe5Ni and (**d**) LSTF-5Fe after calcination, reduced (10 vol.% H_2/Ar, 800 °C, 1 h) and reoxidized (20 vol.% O_2/N_2, 800 °C, 2 h). Features are labelled on the reduced materials according to the underlying Fe-Fe scattering paths.

Figure 11a shows CO conversion curves for the Ni-containing samples LSTN-5Ni, LSTFN-2Fe5Ni and LSTFN-5Fe5Ni. It is apparent that the conversion curves shifted to higher temperatures with increasing Fe concentration, which corresponds to a decrease in catalytic activity. This is in contrast to the previous observation that the presence of Fe increases Ni reducibility (Figure S1). The expected effect would be a larger amount of active Ni^0 and thus higher catalytic activity. On the other hand, the observation is in line with the decrease in WGS activity observed for LSTN-5Fe compared to LSTN in Figure 3b and the indication of Ni/Fe alloy formation during reduction provided by this catalytic activity data, as well as XRD (Figure 5). Since the addition of small quantities of Fe did not appear to be detrimental for catalytic activity, LSTFN-2Fe5Ni was selected for testing catalytic activity with respect to its redox stability, as well as sulfur tolerance. However, it can be seen in Figure 11b that CO conversion decreased over the number of redox cycles, which could be a consequence of Ni/Fe particle growth, due to the incomplete Fe reincorporation over the redox cycles observed by XANES (Figure 9). Despite the fact that Ni re-incorporated completely during reoxidation of reduced

LSTFN-2Fe5Ni (Figure 7), the remaining Fe oxide at the surface may have caused particle growth and thus the observed catalyst deactivation over the consecutive redox cycles.

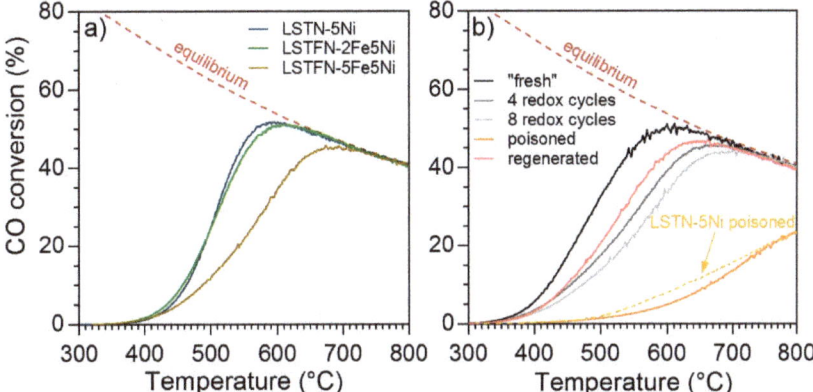

Figure 11. WGS activity of (**a**) reduced LSTN-5Ni, LSTFN-2Fe5Ni, LSTFN-5Fe5Ni (20 vol.% H_2/Ar, 800 °C, 15 h) and (**b**) LSTFN-2Fe5Ni after eight redox cycles, as well as after poisoning with 50 ppm H_2S under reaction conditions (poisoned) and after subjecting the material to one further redox cycle (regenerated). Feed gas composition: 15 vol.% CO/15 vol.% H_2O/7.5 vol.% H_2/Ar; 30,000 mLh^{-1}g^{-1} at STP. The calculated theoretical conversion equilibrium is indicated by the dashed red curve.

The small amounts of Fe (2.5 mol.% or 0.75 wt.%) also did not provide additional stability against poisoning by sulfur, which can be realized from the CO conversion measured after catalyst poisoning by H_2S (Figure 11b). CO conversion was as low as in poisoned LSTN. A complete redox cycle recovered catalytic activity, which was even improved compared to the catalytic activity measured before poisoning. The reason behind this phenomenon was not further investigated as any obvious improvement in performance in terms of the catalytic activity of LSTFN-2Fe5Ni in neither sulfur-free nor sulfur-containing reaction gas feeds could be observed. Nevertheless, it indicates that sulfur can also be successfully removed from LSTFN-type oxides over oxidation-reduction cycles at 800 °C, which could be potentially exploited to completely regenerate these materials if catalyst redox stability can be achieved. The conditions to achieve full reversibility of activity after poisoning should be the aim of future work.

3. Materials and Methods

Mixed metal oxides with nominal composition $La_{0.3}Sr_{0.55}TiO_{3\pm\delta}$ (LST) and $La_{0.3}Sr_{0.55}Ti_{0.95}Ni_{0.05}O_{3\pm\delta}$ (LSTN) were synthesized according to the synthesis procedure described earlier and including a final calcination step at 960 °C (6 h) [27]. Aliquots of these two parent materials were then loaded with Fe, Cr, Mn and Mo precursors by wet impregnation with aqueous precursor solutions followed by drying at 120 °C for around 12 h and subsequent calcination at 500 °C for 2 h. Metal loading was chosen such that the final loading resembled the metal concentration in hypothetical $La_{0.3}Sr_{0.55}Ti_{0.95}Me_{0.05}O_{3\pm\delta}$ and $La_{0.3}Sr_{0.55}Ti_{0.9}Ni_{0.05}Me_{0.05}O_{3\pm\delta}$ (Me = Fe, Cr, Mn and Mo). Table S1 contains all information regarding type and quality of metal precursors used, as well as denotations of the corresponding samples, which are used throughout the text. Furthermore, $La_{0.3}Sr_{0.55}Ti_{0.925}Ni_{0.05}Fe_{0.025}O_{3\pm\delta}$ (LSTFN-2Fe5Ni), $La_{0.3}Sr_{0.55}Ti_{0.9}Ni_{0.05}Fe_{0.05}O_{3\pm\delta}$ (LSTFN-5Fe5Ni), $La_{0.3}Sr_{0.55}Ti_{0.95}Fe_{0.05}O_{3\pm\delta}$ (LSTF-5Fe), and $La_{0.3}Sr_{0.55}Ti_{0.95}Ni_{0.05}O_{3\pm\delta}$ (LSTN-5Ni) powders were also synthesized according to the same procedure, but were calcined at 860 °C for 6 h. After calcination, the powders are referred to as "calcined".

The catalytic activity towards the water gas shift (WGS) reaction was measured on powders in a quartz reactor of plug flow geometry (6 mm ID). Mass flow controllers (Brooks) were used to dose the reactant gases and a K-type thermocouple, which was placed in the middle of the catalyst bed, was used to monitor catalyst bed temperature. To avoid back pressure, all calcined powders were pelletized (4 MPa), crushed and sieved to 100−150 μm before use. The sample (100 mg) was diluted with cordierite powder (200 mg, 75−100 μm) to achieve a thoroughly mixed catalyst bed of ca. 15 mm in length. Catalytic tests were conducted on pre-reduced samples (20 vol.% H_2/Ar, 800 °C, 1 h) after an initial pretreatment of a single redox cycle. This treatment was found to activate LSTN [2] and has therefore been adopted in this work. Catalytic activities were measured at a weight hourly space velocity (WHGS) of 15,000 mL·g^{-1}·h^{-1} at STP (200 mg catalyst, 50 mL·min^{-1}). A Pfeiffer OmniStar GSD 320 quadrupole mass spectrometer equipped with a heated stainless steel capillary was used for compositional analysis of the exhaust gas. CO conversion (X_{CO}) was calculated using Equation (1),

$$X_{CO} = \frac{[CO]^{in} - [CO]^{out}}{[CO]^{in}} \times 100\% \qquad (1)$$

where $[CO]^{in}$ is the initial concentration of CO and $[CO]^{out}$ is the concentrations of CO at the reactor outlet.

Sulfur loading of the catalyst samples was conducted under reaction conditions using the reaction gas mixture (15 vol.% H_2O/15 vol.% CO/7.5 vol.% H_2/Ar), including 50 ppm H_2S. Sulfur loading was conducted at 800 °C for 60 min during which H_2S concentration in the reactor exhaust was monitored using the mass spectrometer signal at M/Z = 34.

CO conversions of LSTN-5Ni, LSTFN-2Fe5Ni, LSTFN-5Fe5Ni were determined on powder samples after an initial activating redox cycle at WGHS= 30,000 mL·g^{-1}·h^{-1} (at STP).

The crystal structure of the powder catalysts was investigated by powder X-ray diffraction (XRD, Bruker D8 Advance) equipped with Ni-filtered Cu Kα-radiation, variable slits and an energy sensitive line detector (LynxEye). Diffractograms were collected at an acquisition time of 4 s and a step size of Δ2θ = 0.03° between 15° and 80°. Aliquots of the samples listed in Table S1 were reduced (10 vol.% H_2/Ar, 800 °C, 1 h) prior to XRD analysis. After reduction, samples were cooled down in Ar (20 °C·min^{-1}). XRD was also recorded on reduced samples at an increased resolution in the angular range 40°–50° (step size 0.005°). The XRD was also recorded for the Ni and Fe containing perovskite-type oxides LSTN-5Ni, LSTFN-2Fe5Ni, LSTFN-5Fe5Ni and LSTF-5Fe after prolonged reduction (10 vol.% H_2/Ar, 800 °C, 15 h).

Temperature programmed reduction (TPR) experiments were conducted using a bench top TPDRO-1100 (ThermoElectron) instrument equipped with mass flow controllers and a thermal conductivity detector. The samples (100 mg) were loaded into the quartz reactor tube and heated to 500 °C under a constant flow of 20 vol.% O_2 before cooling to room temperature. TPRs were recorded in 10 vol.% H_2/Ar (20 mL·min^{-1} at STP) and at a heating rate of 5 °C·min^{-1}. The reoxidation temperature at which Ni is reversibly reincorporated into the perovskite lattice was estimated by TPR redox experiments. A TPR profile was recorded on the calcined sample up to 800 °C followed by an isothermal reduction for 1 h at the same temperature. The sample was then cooled in Ar to room temperature (25 °C) before reoxidation at 700 °C in 20 vol.% O_2/N_2 for 2 h. The sample was again cooled in Ar to 25 °C before starting the second TPR on the now reoxidised material. Such TPR-reduction-reoxidation-TPR cycles were repeated five times with increasing reoxidation temperature (700 °C, 750 °C, 800 °C, 850 °C and 900 °C). The heating rate during reoxidation and cooling after all experiments was 10 °C·min^{-1}.

Ni K-edge (8.333 keV) and Fe K-edge (7.112 keV) X-ray absorption spectra were acquired ex situ on pelletized samples in fluorescence mode at the X10DA (SuperXAS) beamline of the Swiss Synchrotron Light Source (SLS, Villigen, Switzerland) using a 5 element SD detector. The required X-ray energies were scanned using a Si(111) monochromator. The Demeter software package (version

0.9.24) [28] was used to reduce and model all data. The radial distribution function (R) was obtained by Fourier transforming k^3-weighted k-functions typically in the range of 3.0–12.0 Å^{-1} using a Hanning window function. NiO (99.99% trace metals basis, Sigma, Buchs, Switzerland), FeO (99.7% trace metal basis, Sigma), Fe_2O_3 (puriss. ≥97%, Sigma, Buchs, Switzerland), Fe_3O_4 (99.99% trace metal basis, Sigma, Buchs, Switzerland), Fe foil and Ni foil references were measured in transmission mode using ionization chamber detectors. Spectra were recorded on calcined powder samples, after pre-reduction (10 vol.% H_2/Ar, 800 °C, 15 h), as well as after reoxidation (20 vol.% O_2/Ar, 800 °C, 2 h).

Linear combination fitting (LCF) of Ni K-edge X-ray absorption near edge structure (XANES) spectra were performed in the spectral range -20 eV $< E_0 < 30$ eV around the absorption edge to quantify the fraction of each Ni species present in the samples. Reference compounds for each fit included Ni foil, NiO and calcined $La_{0.3}Sr_{0.55}Ti_{0.95}Ni_{0.05}O_{3\pm\delta}$ representing Ni^{n+}_{oct} (n > 2) in the perovskite coordination.

4. Conclusions

Four transition metals (Cr, Fe, Mn and Mo) were screened for potential benefits towards the activity for the water gas shift (WGS) reaction and the stability improvement against sulfur poisoning of $La_{0.3}Sr_{0.55}TiO_{3\pm\delta}$ (LST) and $La_{0.3}Sr_{0.55}Ti_{0.95}Ni_{0.05}O_{3\pm\delta}$ (LSTN). While Cr, Mn and Mo impregnation on LST did not result in active catalysts, Fe exhibited significant WGS activity. Impregnation with Ni produced the most active catalysts. All other metals decreased the intrinsic activity of LSTN suggesting the presence of Ni/metal interactions. Sulfur stability compared to LSTN was improved only in the case of Fe-impregnated LSTN.

Implementing structural reversibility of Fe and Ni was attempted and both metals enhanced reciprocally their reduction behavior. Catalyst oxidation at 800 °C led to complete incorporation of Ni into the host perovskite, whereas Fe incorporation was found to be incomplete under these conditions and resulted in decreased catalyst redox stability at this temperature. Furthermore, WGS activity in the absence of sulfur was reduced compared to LSTN.

Although no beneficial consequences of bimetallic particle segregation were observed, the data demonstrate that Ni catalyst properties towards the WGS reaction at SOFC operation temperatures may be influenced significantly by the presence of other transition metals and that more than one metal can be segregated from LST-type host perovskites. However, the reincorporation behavior may be different for each metal, which has to be taken into account to exploit full structural reversibility of complex systems. Hence, further work is required to optimize regeneration conditions and to exploit the full potential of such materials.

Supplementary Materials: The following are available online at http://www.mdpi.com/2073-4344/9/4/332/s1. Figure S1: Ni speciation from a fit of XANES spectra, Figure S2: Ni K-edge XANES linear combination fit results, Figure S3: k^3-weighted $\chi(k)$ functions at the Ni K-edge, Figure S4: k^3-weighted $\chi(k)$ functions at the Fe K-edge, Table S1: sample list.

Author Contributions: All authors were involved in the conceptualization of this work. P.S. carried out the experiments, analyzed and interpreted the data and wrote the manuscript. D.B. synthesized and provided the perovskite-type materials. O.K. discussed results. A.H. and D.F. analyzed and interpreted the data and served as project leaders. D.F. contributed to writing the manuscript.

Funding: This research was funded by the Competence Center for Energy and Mobility (CCEM) and the Swiss National Science Foundation (SNF, No. 200021_159568).

Acknowledgments: The work was financially supported by the Competence Center for Energy and Mobility (CCEM), the Swiss National Science Foundation (SNF) and the Swiss Federal Office of Energy (SFOE). The work was conducted in the context of the Swiss Competence Center for Energy Research (SCCER BIOSWEET) of the Swiss innovation agency Innosuisse. The X10DA (SuperXAS) beamline at the Swiss Light Source (SLS) in Villigen (Switzerland) and M. Nachtegaal are thanked for kindly providing the beam time and support during measurements.

Conflicts of Interest: The authors declare no conflict of interest.

References

1. Nishihata, Y.; Mizuki, J.; Akao, T.; Tanaka, H.; Uenishi, M.; Kimura, M.; Okamoto, T.; Hamada, N. Self-regeneration of a Pd-perovskite catalyst for automotive emissions control. *Nature* **2002**, *418*, 164–167. [CrossRef]
2. Steiger, P.; Burnat, D.; Madi, H.; Mai, A.; Holzer, A.; Van Herle, J.; Kröcher, O.; Heel, A.; Ferri, D. Sulfur poisoning recovery on a solid oxide fuel cell anode material through reversible segregation of nickel. *Chem. Mater.* **2019**, *31*, 748–758. [CrossRef]
3. Steiger, P.; Nachtegaal, M.; Kröcher, O.; Ferri, D. Reversible segregation of Ni in $LaFe_{0.8}Ni_{0.2}O_3$ during coke removal. *ChemCatChem* **2018**, *10*, 4456–4464. [CrossRef]
4. Atkinson, A.; Barnett, S.; Gorte, R.J.; Irvine, J.T.S.; Mcevoy, A.J.; Mogensen, M.; Singhal, S.C.; Vohs, J. Advanced anodes for high-temperature fuel cells. *Nat. Mater.* **2004**, *3*, 17–27. [CrossRef]
5. McIntosh, S.; Gorte, R.J. Direct hydrocarbon solid oxide fuel cells. *Chem. Rev.* **2004**, *104*, 4845–4865. [CrossRef]
6. Granger, P.; Parvulescu, V.I. Catalytic NO_x abatement systems for mobile sources: From three-way to lean burn after-treatment technologies. *Chem. Rev.* **2011**, *111*, 3155–3207. [CrossRef]
7. Rostrup-Nielsen, J.R.; Pedersen, K. Sulfur poisoning of Boudouard and methanation reactions on nickel catalysts. *J. Catal.* **1979**, *59*, 395–404. [CrossRef]
8. Sehested, J. Four challenges for nickel steam-reforming catalysts. *Catal. Today* **2006**, *111*, 103–110. [CrossRef]
9. Strohm, J.J.; Zheng, J.; Song, C.S. Low-temperature steam reforming of jet fuel in the absence and presence of sulfur over Rh and Rh-Ni catalysts for fuel cells. *J. Catal.* **2006**, *238*, 309–320. [CrossRef]
10. Jahangiri, H.; Bennett, J.; Mahjoubi, P.; Wilson, K.; Gu, S. A review of advanced catalyst development for Fischer-Tropsch synthesis of hydrocarbons from biomass derived syn-gas. *Catal. Sci. Technol.* **2014**, *4*, 2210–2229. [CrossRef]
11. Quinn, R.; Dahl, T.A.; Toseland, B.A. An evaluation of synthesis gas contaminants as methanol synthesis catalyst poisons. *Appl. Catal. A* **2004**, *272*, 61–68. [CrossRef]
12. Boldrin, P.; Ruiz-Trejo, E.; Mermelstein, J.; Menendez, J.M.B.; Reina, T.R.; Brandon, N.P. Strategies for carbon and sulfur tolerant solid oxide fuel cell materials, incorporating lessons from heterogeneous catalysis. *Chem. Rev.* **2016**, *116*, 13633–13684. [CrossRef]
13. Wang, L.S.; Murata, K.; Inaba, M. Development of novel highly active and sulphur-tolerant catalysts for steam reforming of liquid hydrocarbons to produce hydrogen. *Appl. Catal. A* **2004**, *257*, 43–47. [CrossRef]
14. González, M.G.; Ponzi, E.N.; Ferretti, O.A.; Quincoces, C.E.; Marecot, P.; Barbier, J. Studies on H_2S adsorption and carbon deposition over $Mo-Ni/Al_2O_3$ catalysts. *Adsorpt. Sci. Technol.* **2000**, *18*, 541–550. [CrossRef]
15. Murata, K.; Saito, M.; Inaba, M.; Takahara, I. Hydrogen production by autothermal reforming of sulfur-containing hydrocarbons over re-modified $Ni/Sr/ZrO_2$ catalysts. *Appl. Catal. B* **2007**, *70*, 509–514. [CrossRef]
16. Rodriguez, J.A.; Hrbek, J. Interaction of sulfur with well-defined metal and oxide surfaces: Unraveling the mysteries behind catalyst poisoning and desulfurization. *Acc. Chem. Res.* **1999**, *32*, 719–728. [CrossRef]
17. L'Argentière, P.C.; Canon, M.G.; Fígoli, N.S. XPS studies of the effect of Mn on Pd/Al_2O_3. *Appl. Surf. Sci.* **1995**, *89*, 63–68. [CrossRef]
18. Rodriguez, J.A.; Goodman, D.W. The nature of the metal-metal bond in bimetallic surfaces. *Science* **1992**, *257*, 897–903. [CrossRef]
19. Newsome, D.S. The water-gas shift reaction. *Catal. Rev.* **1980**, *21*, 275–318. [CrossRef]
20. Howald, R.A. The thermodynamics of tetrataenite and awaruite: A review of the Fe-Ni phase diagram. *Metall. Mater. Trans. A* **2003**, *34a*, 1759–1769. [CrossRef]
21. Massalski, T.B.; Perepezko, J.H.; Jaklovsky, J. Microstructural study of massive transformations in Fe-Ni system. *Mater. Sci. Eng.* **1975**, *18*, 193–198. [CrossRef]
22. Ratnasamy, C.; Wagner, J.P. Water-gas-shift catalysis. *Catal. Rev.* **2009**, *51*, 325–440. [CrossRef]
23. Steiger, P.; Delmelle, R.; Foppiano, D.; Holzer, L.; Heel, A.; Nachtegaal, M.; Kröcher, O.; Ferri, D. Structural reversibility and nickel particle stability in lanthanum iron nickel perovskite-type catalysts. *ChemSusChem* **2017**, *10*, 2505–2517. [CrossRef]
24. Steiger, P.; Alxneit, I.; Ferri, D. Nickel incorporation in perovskite-type metal oxides – implications on reducibility. *Acta Mater.* **2019**, *164*, 568–576. [CrossRef]

25. Sluchinskaya, I.A.; Lebedev, A.I.; Erko, A. Structural position and charge state of nickel in SrTiO$_3$. *Phys. Solid State* **2014**, *56*, 449–455. [CrossRef]
26. Beale, A.M.; Paul, M.; Sankar, G.; Oldman, R.J.; Catlow, C.R.A.; French, S.; Fowles, M. Combined experimental and computational modelling studies of the solubility of nickel in strontium titanate. *J. Mater. Chem.* **2009**, *19*, 4391–4400. [CrossRef]
27. Burnat, D.; Kontic, R.; Holzer, L.; Steiger, P.; Ferri, D.; Heel, A. Smart material concept: Reversible microstructural self-regeneration for catalytic applications. *J. Mater. Chem. A* **2016**, *4*, 11939–11948. [CrossRef]
28. Ravel, B.; Newville, M. Athena, Artemis, Hephaestus: Data analysis for X-ray absorption spectroscopy using ifeffit. *J. Synchrotron Radiat.* **2005**, *12*, 537–541. [CrossRef]

© 2019 by the authors. Licensee MDPI, Basel, Switzerland. This article is an open access article distributed under the terms and conditions of the Creative Commons Attribution (CC BY) license (http://creativecommons.org/licenses/by/4.0/).

Article

Perovskite-type LaFeO₃: Photoelectrochemical Properties and Photocatalytic Degradation of Organic Pollutants Under Visible Light Irradiation

Mohammed Ismael and Michael Wark *

Institute of Chemistry, Chemical Technology 1, Carl von Ossietzky University Oldenburg, Carl-von-Ossietzky-Str. 9-11, 26129 Oldenburg, Germany; mohammed.ismael1980@gmail.com
* Correspondence: michael.wark@uni-oldenburg.de; Tel.: +49-441-798-3675

Received: 15 February 2019; Accepted: 3 April 2019; Published: 8 April 2019

Abstract: Perovskite-type oxides lanthanum ferrite (LaFeO$_3$) photocatalysts were successfully prepared by a facile and cost-effective sol-gel method using La(NO)$_3$ and Fe(NO)$_3$ as metal ion precursors and citric acid as a complexing agent at different calcination temperatures. The properties of the resulting LaFeO$_3$ samples were characterized by powder X-ray diffraction (XRD), energy dispersive X-ray spectroscopy (EDXS), UV-Vis diffuse reflectance spectroscopy (DRS), X-ray photoelectron spectroscopy (XPS), Fourier transform infrared spectra (IR), transmission electron microscopy (TEM), N$_2$ adsorption/desorption and photoelectrochemical tests. The photoactivity of the LaFeO$_3$ samples was tested by monitoring the photocatalytic degradation of Rhodamine B (RhB) and 4-chlorophenol (4-CP) under visible light irradiation, the highest photocatalytic activity was found for LaFeO$_3$ calcined at 700 °C, which attributed to the relatively highest surface area (10.6 m^2/g). In addition, it was found from trapping experiments that the reactive species for degradation were superoxide radical ions (O$_2^-$) and holes (h$^+$). Photocurrent measurements and electrochemical impedance spectroscopy (EIS) proved the higher photo-induced charge carrier transfer and separation efficiency of the LaFeO$_3$ sample calcined at 700 °C compared to that that calcined at 900 °C. Band positions of LaFeO$_3$ were estimated using the Mott-Schottky plots, which showed that H$_2$ evolution was not likely.

Keywords: sol-gel method; LaFeO$_3$; visible light photocatalysis; perovskite-type structure; Mott-Schottky plot

1. Introduction

Semiconductor-based catalysis is a green technology, which gained considerable attention owing to its potential environmental applications, such as wastewater treatment, air purification and degradation of different organic contaminants [1–3]. In the past few decades, titanium dioxide (TiO$_2$) as an n-type semiconductor has an attractive extensive interest as photocatalysts because of its easy availability, inertness, low costs, nontoxicity and chemical stability [4]. However, the large band gap energy for TiO$_2$ (3.0–3.2 eV), requiring UV light that occupies around 5% of solar energy for excitation, limits its applications to a great extent [5]. Another difficulty is the high recombination rate of the photoexcited electron-hole pairs in TiO$_2$ [6]. Many attempts have been developed to retard this electron-hole recombination and to increase the photocatalytic efficiency of TiO$_2$, such as surface modification using a suitable metal ion and nonmetal dopant to increase the visible light absorbance and coupling with another semiconductor to enhance the charge separation efficiency [7,8]. Although in some cases improved photocatalytic activities were reported, very often the doping increased the number of structural defects acting as unwanted recombination centers. Therefore, the development of cost-effective, efficient and alternative photocatalysts with intrinsic narrow band gaps to increase the visible light response has become a research focus [9,10]. Mixed metal oxides and oxynitrides attracted interest since many of them are visible-light active, cheap, non-toxic and stable [11].

Iron is highly abundant in the earth crust and thus cheap. Many mixed metal oxides containing iron, i.e., ferrites, offer suitable band gap energy for visible light absorption. Furthermore, the position of their valence band edges is more positive than the oxidation potential of O_2/H_2O (1.23 V vs. NHE) rendering them suitable for the photooxidation of water [12]. The high activity of ferrites for degradation of pollutants has been proven in many studies [13–15]. Ferrites with a perovskite structure, with a general formula of ABO_3 with for example, A = rare-earth metal ion and B = Fe^{3+} ion, exhibit a wide range of ferro-, piezo-, and pyro-electrical properties rendering them suitable as magneto-optical material, electrode materials, structural materials, sensors and refractory materials [16]. The perovskite $LaFeO_3$ is employed as a catalyst, e.g., in solid oxide fuel cells, but also in devices using its good dielectric properties and high piezoelectricity.

However, $LaFeO_3$ has also been used as a photocatalyst; several studies focused on the synthesis and the activity for photodegradation of several organic dyes under visible light irradiation [17–19]. Thirumalairajan et al. synthesized floral-like $LaFeO_3$ by a surfactant-assisted hydrothermal technique and found that the porous floral nanostructure led to higher photoactivity compared to bulk $LaFeO_3$ for the degradation of different dyes, such as rhodamine B (RhB) and methylene blue (MB) [20]. Su et al. prepared large surface area nanosized $LaFeO_3$ particles by employing SBA-16 as a hard template and compared its visible light activity for RhB degradation with that of $LaFeO_3$ prepared by the citric acid assisted sol-gel route [21]. Yang et al. prepared $LaFeO_3$ by conventional co-precipitation and enhanced its activity by post-treatment in molten salt [22]. Tijare et al. [23] formed nano-crystalline $LaFeO_3$ perovskite by the sol-gel route and claimed activity for photocatalytic hydrogen generation under visible light irradiation.

In the present work, we applied the same synthetic route for $LaFeO_3$ as Tijare et al. but altered (i) the duration of the thermal treatment and (ii) used a pyrolysis step at 400 °C instead of using ultra-sonication or drying at 90 °C in an oven. Citric acid assisted sol-gel was chosen as a synthesis route because, in general, it is a suitable method for the synthesis of nanopowders with a well-developed high specific surface area obtained at low calcination temperature and short times without employing expensive sacrificial structure-directing agents or template structures. The visible light activity for degradation of RhB and 4-chlorophenol (4-CP) as model organic pollutants was investigated. As for Tijare et al., we also attempted hydrogen generation, however, failed with that and suspected it was based on the Mott-Schottky plots calculating band positions that the conduction band edge of $LaFeO_3$ was too positive than the reduction potential of H_2/H_2O (0 V vs. NHE) to create electrons which were reductive enough to react with protons to hydrogen.

2. Results and Discussion

2.1. Structural and Optical Characterization of $LaFeO_3$

The powder XRD patterns of the prepared $LaFeO_3$ samples after calcination at various temperatures are shown in Figure 1. All the diffraction peaks belong to the orthorhombic $LaFeO_3$ with ABO_3-type perovskite structure (JCPDS card No. 88-0641) [24]. The main characteristic reflexes are located at 2θ of 22.6°, 25.5°, 32.2°, 34.5°, 39.7°, 46.1°, 47.7°, 52.0°, 54.0°, 57.4°, 64.0°, 67.3° and 76.6°, being indexed to the (101), (111), (121), (210), (220), (202), (230), (141), (240), (115), (242) and (204) diffraction planes, respectively [24]. This confirms the effective preparation of a single phase perovskite $LaFeO_3$ without any crystalline impurities like La_2O_3 or Fe_2O_3. For calculating the crystallite sizes the reflexes of highest intensity at 2θ = 32.2° were selected. With increasing calcination temperature the diffraction peaks get sharper and more intense, indicating a better crystallization and growth of the grains. The average crystallite sizes of $LaFeO_3$ D have been determined by using the Debye-Scherer formula [25]:

$$D = \frac{K\lambda}{\beta cos\theta}$$

with K being the crystallite shape factor, λ the X-ray wavelength (1.5406 nm for Cu Kα), β is the width of the diffraction peak and θ is the Bragg angle. The crystallite sizes were 27.4 nm, and 45.7 nm for S-700 and S-900, respectively. In the smaller particles, less time was needed for the electrons and holes to diffuse from the inner part to the surface of the catalyst, where they could react. This typically leads to higher photocatalytic efficiency.

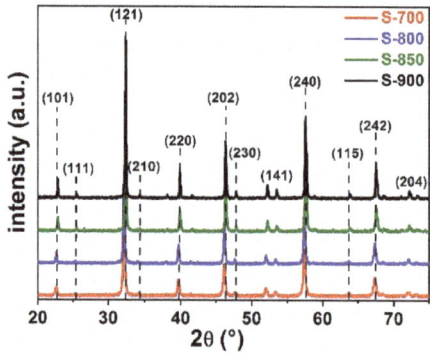

Figure 1. X-ray diffraction patterns of LaFeO$_3$ obtained at different calcination temperatures.

Figure 2 shows the scanning electron micrographs of the prepared LaFeO$_3$ samples at different calcination temperatures. Scanning electron microscopy (SEM) was used to determine the morphology of the perovskite LaFeO$_3$ samples; as seen in Figure 2 both samples show a network structure with semi-spherical morphology. It was found that the particle sizes of S-700 were significantly smaller than those of S-900, consistent with the trend of the crystallite sizes determined from XRD.

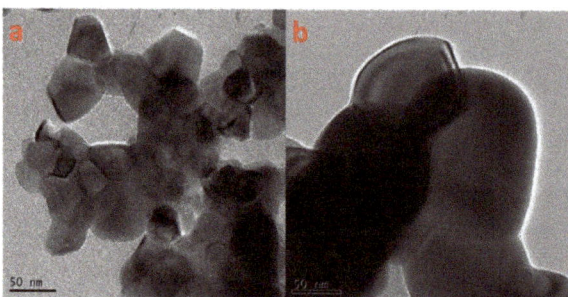

Figure 2. SEM images of LaFeO$_3$ calcined at different temperatures: (**a**) S-700, (**b**) S-900.

Energy dispersive X-ray spectroscopy (EDXS) was used to investigate the purity and chemical composition of synthesized LaFeO$_3$ nanoparticles, the pattern of calcined LaFeO$_3$ samples are shown in Figure 3. Besides a carbon signal appearing at 0.277 keV and resulting from the latex of the SEM sample holder due to the incomplete coverage of the sample [25], only lanthanum (La), iron (Fe) and oxygen (O) were present, confirming that the citric acid assisted sol-gel route leads to high purity LaFeO$_3$ photocatalyst. The peaks at around 0.83 and 4.65 keV were related to La and the ones at around 6.399 and 0.704 keV to Fe; they proved that the formation of the LaFeO$_3$ photocatalyst had a 1:1 molar ratio of metal ions as the atomic percentage obtained from EDXS was 0.35% for Fe and 0.34% for La in sample S-700 and 1.13% for Fe and 1.21% for La in S-900, respectively.

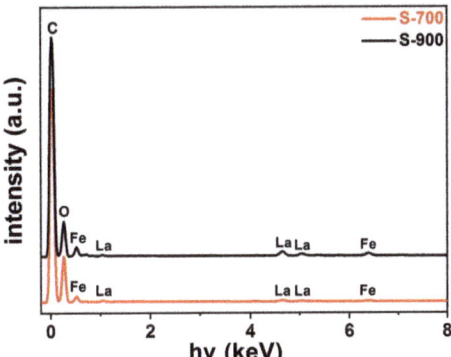

Figure 3. Energy dispersive X-ray (EDX) spectra of LaFeO$_3$ at different calcination temperatures (S-700, S-900).

Figure 4 presents the IR spectra of different LaFeO$_3$ samples prepared at different calcination temperatures in the wavenumber range 400–4000 cm^{-1} in order to determine the possible functional groups in the sample. The FT-IR spectra were quite featureless, confirming again the purity of the synthesized samples. The peak at 556 cm^{-1} can be attributed to the Fe–O stretching vibration being characteristic of the FeO$_6$ octahedrons in perovskite-type LaFeO$_3$ [26]. The band at 716 cm^{-1} can be assigned bending vibrations of the La-O bonds [11]; the small peak at 2905 cm^{-1} as well as the small peak at around 1600 cm^{-1} resulted from small amounts of citric acid residues, they were accounted to C-H vibrations and the symmetric stretching of the carboxyl groups, respectively.

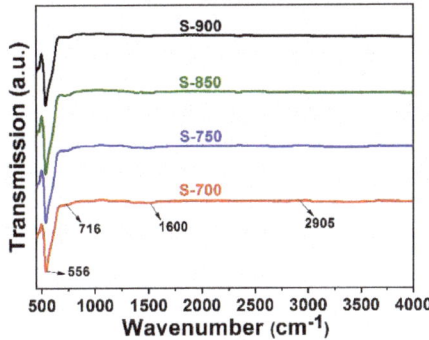

Figure 4. FT-IR spectra of LaFeO$_3$ nanoparticles calcined at different temperatures.

The specific surface areas of the synthesized LaFeO$_3$ samples were determined from a nitrogen adsorption-desorption isotherm using the BET approach. The isotherms of the samples can be classified into type III behavior (Figure 5), which is attributed to a weak adsorbate-adsorbent interaction [27]. The surface areas decreased with increasing calcination temperature. Although the surface areas were in general quite small, the area of S-700 exceeded even slightly the highest value reported by Tijare et al. of 9.5 m^2/g [23]. In general, higher surface areas facilitated adsorption of organic pollutants, promoted charge carrier separation and enabled more light harvesting, resulting in total higher photocatalytic activity.

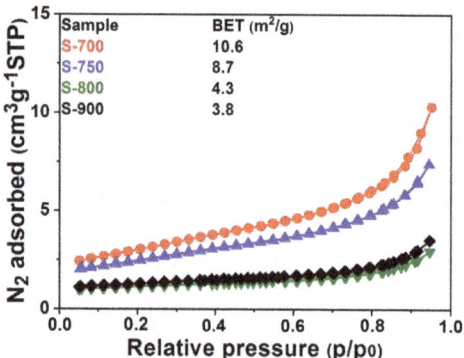

Figure 5. N$_2$ adsorption-desorption isotherms for LaFeO$_3$ samples and the resulting BET surface areas of the different samples.

Diffuse reflectance UV-Vis spectroscopy was employed to characterize the optical properties of the LaFeO$_3$ nanoparticles, as shown in Figure 6a. In the perovskite-type oxide, the strong absorption edge at 520 nm was ascribed to the electronic transition from the O 2p orbitals forming the valence band to the Fe 3d orbitals in the conduction band [28]. The data showed that the sol-gel prepared LaFeO$_3$ photocatalyst could serve as a potential visible-light-driven photocatalyst. In addition, the band gap energy of LaFeO$_3$ catalysts can be determined from Kubelka−Munk equation [29] via a Tauc plot:

$$\alpha = B\left(h\nu - E_g\right)^n / h\nu$$

with α being the absorption coefficient, ν the irradiation frequency, E_g the band gap, B being a constant (being usually 1 for semiconductors), h is the Planck constant and n is a constant depending on the type of semiconductor (direct transition: n = 1/2; indirect transition: n = 2). For the direct transition semiconductor LaFeO$_3$, the band gap energy values were estimated by extrapolation of the linear part of the curves of the Kubelka–Munk function $(\alpha h\nu)^{1/2}$ against the photon energy ($h\nu$), as displayed in Figure 6b. The S-700 sample absorbed slightly more light energy than the other ones showing that a decreasing particle size and an increased surface area led to a slight red shift.

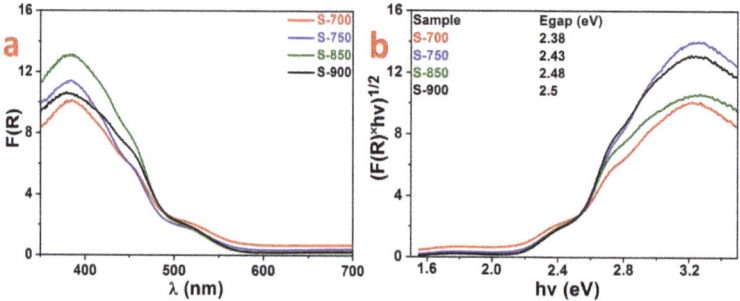

Figure 6. Kubelka-Munk diffuse reflectance UV-Vis spectra (a) and resulting band gaps from Tauc plots (b) of the studied LaFeO$_3$ samples.

X-ray photoelectron spectroscopy (XPS) was performed on the most promising LaFeO$_3$ S-700 to determine the elemental composition and the chemical oxidation state of the sample surface. Figure 7 shows the X-ray photoelectron (XP) survey spectrum (Figure 7a) and the detailed spectra of the La 3d, Fe 2p and O 1s. In addition to different La, Fe and O lines, Figure 7a shows also the C 1s signal, which resulted from adventitious surface carbon and which was also referenced to a binding energy

(284.8 eV) in order to exclude surface charge effects for all the other signals. The binding energies found in the XP spectra of La 3d (Figure 7b) and Fe 2p (Figure 7c) revealed that the iron and lanthanum ions were both present in the chemical valence state +III [30,31]. Figure 7d shows the O 1s signal with binding energies of about 529.9 eV, 531.1 eV and 532.0 eV which corresponded as the main signal to the contribution of the La-O and Fe-O crystal lattice bonds, some surface hydroxyl groups and chemisorbed water, respectively [32,33]. In line with the expected composition between La, Fe and O, an atomic ratio of about 1:1:3 was found by comparing the relative signal intensities in the different XP spectra.

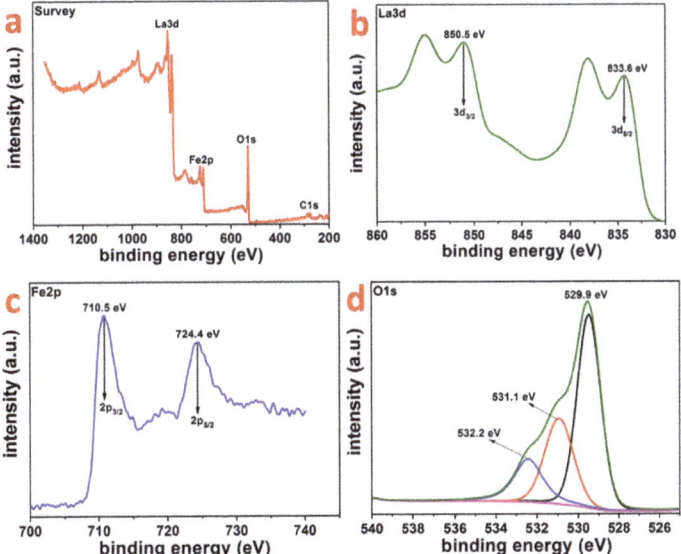

Figure 7. X-ray photoelectron spectroscopy (XPS) for the S-700 sample survey (**a**), and core level XPS of La 3d (**b**), Fe 2p (**c**) and O1s (**d**).

In order to investigate the energy band structure including the conduction band (CB) and valence band (VB) position of LaFeO$_3$ S-700 and S-900 photocatalysts, electrochemical flat potential measurements were performed, and the resulting data are plotted in Figure 8a,b using the Mott–Schottky (MS) relation in the dark [34]:

$$\frac{1}{C^2} = \left[\frac{2}{\varepsilon\varepsilon_o eN}\right]\left[E - E_{fb} - \frac{kT}{e}\right]$$

with C being the capacitance of space charge, ε the dielectric constant, ε_o the permittivity of free space, N the electron donor density, E the applied potential and E_{fb} the flat band potential. Plotting $1/C^2$ against E yields a straight line from which the slope of the donor density can be calculated and E_{fb} can be determined as the intercept of the abscissa by extrapolation to $C = 0$. The positive slopes of the MS plots confirmed LaFeO$_3$ being an n-type semiconductor with electrons as the majority charge carriers. The flat band potentials of S-700 (Figure 8a) and S-900 (Figure 8b) at frequencies of 100 Hz and 1000 Hz were calculated to be -0.3 and -0.25 V ($E_{Ag/AgCl(sat-KCl)}$) referenced to the KCl-saturated Ag/AgCl electrode, respectively. Thus, using the following equation:

$$E_{NHE} = E^0{}_{Ag/AgCl(sat-KCl)} + E_{Ag/AgCl(sat-KCl)} + 0.059 \times pH$$

with $E^0_{Ag/AgCl(sat-KCl)}$ = 0.199. For the pH value of 5.6 of the 0.1 M Na_2SO_4 electrolyte solution, potentials were 0.23 and 0.28 V versus the normal hydrogen electrode (NHE) result. For n-type semiconductors the E_{fb} was strongly related to the bottom of the conduction band (CB); typically, it is assumed that CB is 0.1 V more negative than E_{fb} [35], resulting in CB edges at about 0.13 V and 0.18 V vs. NHE, respectively. These values are close to the position of the conduction band edge for $YFeO_3$ calculated by Ismael et al. [36]. The slightly positive CB potential explains that the $LaFeO_3$ was not able to form hydrogen via water splitting under light irradiation. This was confirmed by respective experiments attempting photocatalytic H_2 formation with our $LaFeO_3$ samples on which platinum nanoparticles (0.5 wt. %, particle size <2 nm) were photodeposited as a co-catalyst. Even by the use of light with λ ≥ 320 nm and methanol as a sacrificial agent, no hydrogen was detected with all the $Pt/LaFeO_3$ samples. This result stands in contrast to H_2 formation reported earlier by Tijare et al. [23], Parida et al. [37] and Vaiano et al. [38], who, however, performed no analysis on conduction band positions. Thus, some doubts regarding the H_2 production reported in their papers exist.

Xu et al. [39] reported hydrogen production activity over a $LaFeO_3/g-C_3N_4$ composite in the presence of TEOA as a sacrificial reagent and Pt as a co-catalyst. Their results show that $LaFeO_3$ alone had no activity due to the positive conduction band edge (0.11 V); hydrogen was only found if $g-C_3N_4$ (conduction band potential at −0.85 V vs. NHE) was added, which is in agreement of our results. Hydrogen production was observed for other ferrites, such as $CuFe_2O_4$ and $NiFe_2O_4$; for those the conduction band positions were found to be negative enough [40,41].

By taking into account the band gap energies of our $LaFeO_3$ samples from the Tauc plots (Figure 6b), the valence band (VB) positions for S-700 and S-900 can be calculated according to the equation $E_{vb} = E_{cb} + E_g$ [42], resulting in about E_{vb} = 2.51 V and 2.68 V respectively.

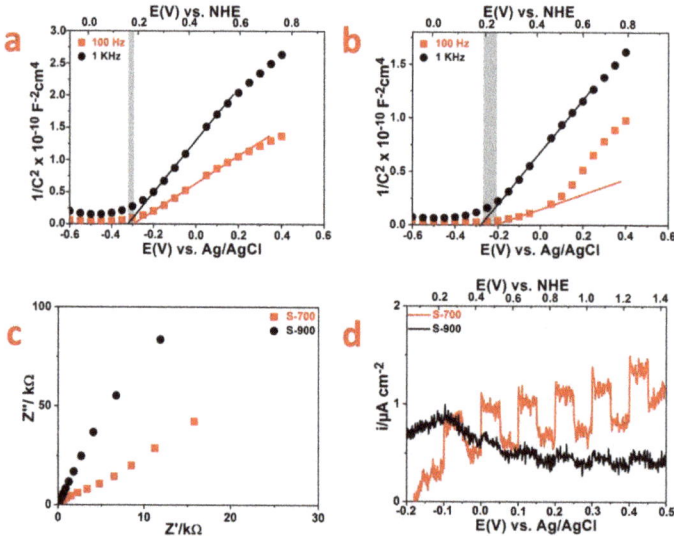

Figure 8. Mott-Schottky plots at 100 Hz and 1 kHz of (**a**) S-700 and (**b**) S-900, (**c**) Nyquist plots of S-700 and S-900 in 0.1 M Na_2SO_4 (pH = 5.6) at 0.4 V vs. Ag/AgCl and (**d**) transient photocurrent responses in 0.1 M Na_2SO_4/0.1 M Na_2SO_3 solution (pH = 5.6) under white LED illumination.

Electrochemical impedance spectroscopy (EIS) and transient photocurrent experiments were performed to investigate the electron-hole separation efficiency in the $LaFeO_3$ photocatalysts. The electrode of S-700 shows the smaller arc size (Figure 8c). In general, a smaller arc size observed in EIS semicircular Nyquist plots documents smaller charge-transfer resistance on the electrode surface and accelerated interface transport of charge carriers, which results in an effective photo-induced charge carrier mobility and separation [43,44]. Figure 8d indicates the transient photocurrent responses of S-700 and S-900. The photocurrent of S-700 sample was much higher than that of S-900 indicating the greatly improved charge transfer and separation ability [45,46]. The onset potential of the photocurrent indicates the flat band potential of the electrode [47]. In this case, the sulfate/sulfite electrolyte solution lowered the kinetic barrier for charge transport by trapping the photogenerated holes. Moreover, the onset of the photocurrent lies at about 0.2 V vs. NHE for S-700, being in good agreement with the flat band potential obtained from the Mott-Schottky plot.

2.2. Photocatalytic Properties

Photocatalytic activities of the prepared LaFeO$_3$ samples were evaluated by degradation of RhB and 4-CP in aqueous solution under visible light irradiation using a 420 nm cut-off filter. Before irradiation, the suspensions were magnetically stirred in the dark for 40 min to ensure adsorption-desorption equilibrium between the organic substrate and the photocatalyst, after visible light irradiation the absorbance of RhB was noticeably reduced (Figure 9a), although there was very little decrease in absorption before irradiation. This indicates that RhB degradation occurred instead of further adsorption. Since the intensity of the absorption peaks gradually decreased without any change in their wavelength, it can be concluded that the degradation reaction takes place by an aromatic ring opening without formation of stable de-ethylated intermediates [48,49].

Figure 9b shows that without LaFeO$_3$ being present the dye RhB was quite stable and no significant self-degradation under visible light took place. Also in the presence of SnO$_2$, a semiconductor with a band gap of 3.0 eV, which can, thus, not be excited by light with $\lambda \geq 420$ nm, only negligible degradation was found. Thus, sensitization effects can be ruled out as well. In the presence of the photocatalyst LaFeO$_3$, the photodegradation efficiency decreased with increasing calcination temperature of the LaFeO$_3$ due to the decreasing surface area and increasing particle size. Besides the highest surface area sample facilitating the adsorption of the organic dyes and possibly trapping more electrons and holes on the surface, the sample S-700 might also benefit from the slightly narrower optical band gap allowing for more visible light absorption.

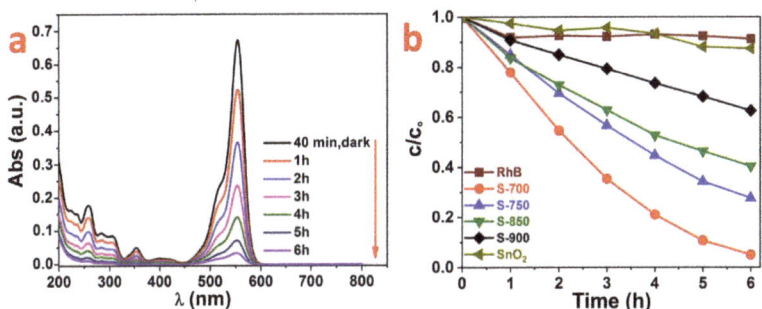

Figure 9. (a) UV-Vis spectra for the degradation of rhodamine B (RhB) under visible light irradiation ($\lambda \geq 420$ nm) on the LaFeO$_3$ S-700 sample ([RhB] = 10^{-5} M, catalyst weight = 0.1 g) at 25 °C. (b) Degradation of RhB as a function of irradiation time with visible light ($\lambda \geq 420$ nm) in the presence of different LaFeO$_3$ catalysts or SnO$_2$ for comparison.

The degradation of RhB obeys a pseudo-first-order kinetics law of the type [50]:

$$\ln c_0/c_t = kt$$

with c_0 being the initial concentration of RhB, c_t the concentration of RhB at any time t, t the illumination time (min) and k is the first order rate constant (min^{-1}). Figure 10 shows the linear relationships between $\ln(c_0/c_t)$ and t, the rate constant for S-700, S-750, S-850 and S-900 were 0.0062, 0.0032, 0.0026 and 0.0013 min^{-1}, respectively.

In general, it is known that the photocatalytic degradation reaction of organic contaminants proceeds mainly by the contribution of oxygen-containing reactive species such as superoxide ($\cdot O_2^-$) and hydroxyl radicals ($\cdot OH$) [51,52]. Thus, in order to explore the reactive species for RhB degradation different scavengers were tested in the photocatalytic process. The hydroxylation test was done using terephthalic acid (TA) as a probe molecule, in this test TA reacts with OH to produce highly fluorescent 2-hydroxyterepthalic acid (fluorescence maximum at 426 nm [53]).

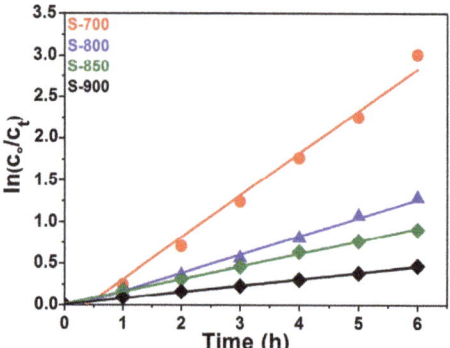

Figure 10. Kinetics curves of LaFeO$_3$ perovskites calcined at different temperatures.

As seen from Figure 11a for S-700 very low fluorescence intensity at 426 nm was observed after 6 h of visible light irradiation suggesting very low hydroxyl radical formation on the surface of the catalyst. Usually, the photoluminescence (PL) intensity at about 425 nm is proportional to the amount of the produced hydroxyl radical on the surface of the catalyst. In good agreement to that, isopropanol (0.01 M), a known hydroxyl radical ($\cdot OH$) quencher [54], showed only little effect on the RhB degradation reaction (Figure 11b). However, the addition of benzoquinone (0.01 M) [55] as a superoxide radical ($\cdot O^-_2$) quencher strongly decreased the degradation of RhB, indicating that degradation proceeds via superoxide radicals, which can only be produced via the reduction of dissolved oxygen by the excited electrons in the conduction band (CB) of LaFeO$_3$. This is surprising since the CB of LaFeO$_3$ S-700 was detected to be at about 0.1 V, being more positive than the standard potential for the superoxide radical formation from adsorbed oxygen $E^0(O_2/\bullet O_2^-) = -0.046$ V [56]. Thus, the superoxide radical formation should not be possible. However, in the photocatalytic experiment, the electrochemical standard conditions were not given, thus due to potential shifts depending on the Nernst law, the superoxide radical formation might become possible to some extent. An indication for potential changes during the photocatalysis experiment might become visible in the increasing degradation of RhB in the presence of benzoquinone, which occurs with longer irradiation time (Figure 11b). Approximately the same decrease in activity for RhB degradation is obtained when 10 vol. % methanol [57] were added as a hole (h$^+$) quencher. The photogenerated holes can oxidize OH$^-$ ions to $\cdot OH$ radicals because the valence band position of S-700 (2.46 V) is more positive than the redox potential of $\cdot OH/^-OH$ (E^0 = 1.99 V) [58]. These results for the active species being responsible for degradation agree with the results for the perovskite YFeO$_3$ studied earlier by us [36]. In that paper, it was concluded that superoxide radicals ($\cdot O_2$) and ($\cdot OH$) have an effect but the holes (h$^+$) are the main species on catalyst surfaces responsible for the photocatalytic activities.

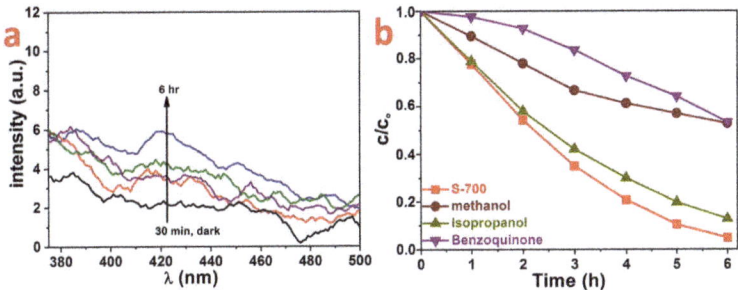

Figure 11. (a) Fluorescence spectra of the 2-hydroxyterepthalic acid solution in the presence of LaFeO$_3$ S-700 and (b) the reactive species in trapping experiments during the degradation of RhB.

Figure 12 shows that the LaFeO$_3$ S-700 was not only able to degrade dyes like RhB under visible-light irradiation at λ > 420 nm, but also compounds like 4-CP which does not absorb lights themselves in that spectral range. Thus, a light-induced self-degradation can be ruled out. With LaFeO$_3$ S-700 the degradation efficiency on 4-CP was lower than that on RhB, but still more than 60% of the 4-CP were transformed by destroying the aromatic ring system, which is responsible for the absorption at 315 nm, which was recorded and followed with time in Figure 12. In a HPLC analysis performed with the reaction mixture after 5 h of illumination, no significant amounts of decomposition products were found. Total organic content (TOC) analysis after five hours illumination time confirmed the 4-CP degradation; a reduction of the TOC by 55% was found. Our LaFeO$_3$ photocatalyst showed higher activity for 4-CP degradation compared to that reported by Pirzada et al. [59] and Hu et al. [60].

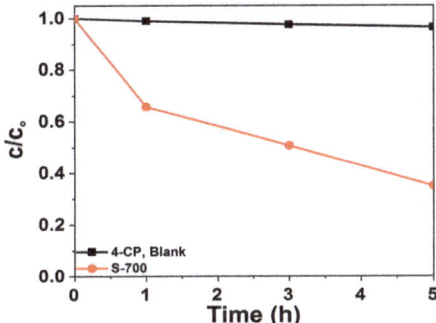

Figure 12. Degradation of 4-CP as a function of irradiation time without a photocatalyst and in the presence of the LaFeO$_3$ catalyst S-700.

3. Experimental

3.1. Materials

All the reactants were commercial products. Iron (III) nitrate nonahydrate (Fe(NO$_3$)$_3$·9H$_2$O, 99%), citric acid (C$_6$H$_8$O$_7$, 99%) and lanthanum (III) nitrate hexahydrate (La(NO$_3$)$_3$·6H$_2$O, 99.999%) were purchased from Sigma-Aldrich. 4-chlorophenol (4-CP, ClC$_6$H$_4$OH, ≥99%) and Rhodamine B (RhB, C$_{28}$H$_{31}$ClN$_2$O$_3$) were bought from Merck. All chemicals were analytically pure and were used as received without further purification. Deionized water (>18 MΩ cm) obtained from a Sartorius water purification system was used during the whole synthetic procedure to prepare the required metal ion solutions.

3.2. Synthesis of LaFeO₃ by the Citric Acid Assisted Sol-gel Method

The lanthanum ferrite perovskite was synthesized by the citric acid assisted sol-gel method [23]. In a typical synthesis, $Fe(NO_3)_3 \cdot 9H_2O$ (0.0041 moles), $La(NO_3)_3 \cdot 6H_2O$ (0.0041 moles) and citric acid were separately dissolved in deionized water under continuous stirring in a 1:1:4 molar ratio. A high surplus of citric acid was required to chelate the metal cations and to prevent aggregation. The obtained dark yellow, clear and transparent solution was obtained after the powders were completely dissolved in the solution. From that, the viscous gel was formed by heating at 300 °C for 2–3 h under continuous magnetic stirring. Subsequently, the solvent was evaporated by combustion in a pyrolysis setup at 400 °C for 1 h. The resulting fluffy powder, which was used as a precursor for $LaFeO_3$, was crushed to a fine powder and subsequently calcined at different temperatures for 4 h in air. After cooling, the obtained samples were characterized and tested in the photocatalytic degradation. They were named S-700, S-750, S-850 and S-900 according to the calcination temperature applied.

3.3. Characterization

The crystalline phase and size of the obtained $LaFeO_3$ nanoparticles were checked with a Empyrean theta-theta X-ray diffraction system (PANalytical, Almelo, The Netherlands) operating with Cu Kα radiation (λ = 1.540598 nm) at 40 kV and 40 mA in the 2θ range of 20–80°. The Brunauer-Emmett-Teller (BET) surface areas were calculated from nitrogen adsorption-desorption isotherms measured on a Tri Star II (Micromeritics GmbH, Aachen, Germany). All the samples were degassed at 150 °C overnight prior to the adsorption measurements.. The diffuse reflectance spectra (DRS) and UV-Vis absorption spectra of the dye solutions were measured with a Varian Cary 4000 (Mulgrave, Australia). This spectrometer could also be equipped with an Ulbricht sphere allowing the recording of UV-Vis diffuse reflectance spectra (DRS) in the region 200–800 nm using the white standard MgO as a reference. The photoluminescence (PL) spectra were recorded using a Varian Cary Eclipse fluorescence spectrophotometer (Mulgrave, Australia) at room temperature, with excitation by incident light of 380 nm. The Fourier transform infrared spectra (FT-IR) were recorded using a Bruker FT-IR Tensor 27 Spectrometer with a platinum ATR unit. The morphology of the prepared samples and the content of the elements were studied using scanning electron microscopy (Hitachi, S-3200N, Krefeld, Germany) and energy-dispersive X-ray spectroscopy (EDX Oxford INCAx-act, Abingdon, UK), respectively. Since the photocatalytic properties of the $LaFeO_3$ strongly depend on the surface chemical state of the samples this was analyzed by X-ray photoelectron spectroscopy (XPS) with an ESCALAB 250 Xi (Thermo Fisher, East Grinstead, UK) equipped with a monochromatic Al Kα X-ray source (hν = 1486.6 eV) as the excitation source under ultrahigh vacuum conditions. The high-resolution spectra for the C 1s, O 1s, La 3d and Fe 2p photoelectron lines were recorded with a bandpass energy of 20 eV and a step size of 0.1 eV. The spectra were analyzed using Avantage software (version 5.951). Binding energies in the high-resolution spectra were calibrated by setting the C 1s signal to 284.8 eV. Electrochemical impedance spectroscopy (EIS), photocurrent measurement and Mott Schottky plots were performed on an IM6e potentiostat (Zahner Elektrik, GmbH, Kronach, Germany) and evaluated with the Thales (version 4.12) software. A standard three-electrode cell configuration with Ag/AgCl (saturated KCl), coiled Pt-wire and $LaFeO_3$ coated FTO were used as the reference, counter and working electrodes, respectively. The working electrodes were prepared on the conductive fluorine-doped tin oxide (FTO) glass slides (Pilkington, Weiherhammer, Germany) of 2 × 6 cm size, which were sonicated before by applying in a sequence 0.1 M HCl, 0.1 M NaOH, acetone and ethanol in an ultrasonic bath, then rinsed with deionized water and dried in an air stream. Then 0.1 g of the respective $LaFeO_3$ sample was dispersed in 500 µl of ethanol in an ultrasonic bath for 30 min and 200 µl of the obtained suspension were coated on the FTO glass substrate using the doctor blade method. Finally, the FTO substrate was dried in a furnace for 30 min at 353 K and calcined for 3 h at 873 K. An electrolyte 0.1 M Na_2SO_4 aqueous solution was used for the Mott-Schottky measurements and a 0.1 M Na_2SO_4/0.1 M Na_2SO_3 aqueous solution was used as the supporting electrolyte for photocurrent and electrochemical impedance measurements, respectively.

3.4. Photocatalytic Degradation Activity and Hydrogen Evolution Measurements

The photocatalytic activity of the LaFeO$_3$ samples was tested on one hand by the photocatalytic decomposition of RhB and 4-CP as organic model pollutants under visible light irradiation ($\lambda \geq 420$ nm) and on the other hand by attempting hydrogen evolution using light of $\lambda \geq 320$ nm and Pt nanoparticles as a co-catalyst.

For the degradation experiments the visible light irradiation was obtained from a 150 W Xe lamp equipped with a 420 nm cut-off filter. This optical cut-off filter was placed between the reactor and the xenon lamp to cut off the UV light and to ensure visible light irradiation only; the distance between the light source and the reactor containing solution was about 10 cm. Since for dyes, such as RhB, self-excitation by the visible light could not be ruled out (although we found there was not much indication for that, compare Figure 9b), 4-CP was used as a control substrate, which is not absorbing in the vis-light range. In a typical test, 0.1 g of LaFeO$_3$ and 100 mL of aqueous solutions (10^{-5} M) of RhB or 4-CP were initially mixed under continuous magnetic stirring in a water-cooled (10 °C) double-wall 250 mL Pyrex reactor for 40 min in the dark to establish the adsorption-desorption equilibrium between the photocatalyst and substrate (RhB or 4-CP). The cooling system was used to cool down the double-wall Pyrex reactor to prevent the effect of the thermal catalytic reaction. Then the samples were illuminated under visible light. Every one hour, a part of the suspension was taken out, filtered to remove the particles and analyzed by the UV-Vis absorption measurement, the degree of degradation was evaluated from the decrease of absorption at the RhB and 4-CP maxima at 554 nm and 315 nm, respectively.

Photocatalytic H$_2$ evolution was attempted in a double-walled quartz reaction vessel connected to a closed gas circulation, using a 500 W Hg mid-pressure immersion lamp (Peschl UV-Consulting) as a light source. Argon gas was used as the carrier gas with a flow of 50 NmL min^{-1}. The evolved hydrogen gas was quantitatively analyzed by a multichannel analyzer (Emerson) equipped with a thermal conductivity detector. In a typical photocatalytic reaction, 0.5 g of LaFeO$_3$ photocatalysts were suspended in a mixture of 550 mL water and 50 mL of the sacrificial agent methanol prior to the irradiation. The solution was kept at 10 °C by flushing cold water from a thermostat (LAUDA) through a double-wall reactor made of normal glass. The normal glass mantle was tested to absorb all the light with wavelengths shorter than about 320 nm. Thus, one can assume that the LaFeO$_3$ samples are only irradiated by light of $\lambda \geq 320$ nm. The co-catalyst 0.5 wt. % Pt (particle size < 2 nm) was deposited on the LaFeO$_3$ powder via reductive photodeposition from H$_2$PtCl$_6$·6H$_2$O. Upon light irradiation, metallic Pt nanoparticles were photodeposited onto the photocatalyst surface sites preferentially accessible for electrons, while CO$_2$ was formed from methanol being employed as a sacrificial reagent as qualitatively detected with our multichannel analyzer (Emerson). The standard redox potential for the reduction of Pt^{2+} ions to metallic Pt is +1.2 V vs. NHE, thus the electrons in the CB of LaFeO$_3$ are able to initiate this reduction.

4. Conclusions

A perovskite-type LaFeO$_3$ photocatalyst was synthesized using the quite simple citric acid assisted sol-gel route. The prepared samples were characterized using different methods. The most photocatalytically active sample for decomposition of RhB and 4-CP under visible light was the one calcined at the lowest temperature of 700 °C due to the highest surface area and the lowest band gap energy. At temperatures lower than 700 °C the crystallinity of the LaFeO$_3$ samples was not sufficient. Mott-Schottky plots revealed a positive potential of the CB at around 0.1 V explaining the observed inactivity of LaFeO$_3$ for the photocatalytic hydrogen evolution via water splitting and methanol dehydrogenation. The photocatalytic degradation reaction of the pollutants occurred mainly via direct reaction of the photogenerated holes, but to some extent especially in the starting period of the degradation experiments also via superoxide radical formation. To sum up, LaFeO$_3$ is a promising photocatalytic material for degradation of organic pollutants under visible light irradiation.

Author Contributions: M.I. performed the catalyst preparation, structural and electrochemical characterization and the photocatalytic test experiments. He also prepared the first draft of the manuscript. This work was carried out under the supervision of M.W., who also structured the discussion of the data and prepared the final version of the manuscript.

Acknowledgments: We thank Dereje H. Taffa (University of Oldenburg) for his assistance in recording XP spectra and fruitful discussions XPS and photoelectrochemical results. Financial support by the Phoenix Scholarship program (PX14DF0164) for MI, and by the German Science Foundation (Deutsche Forschungsgemeinschaft, DFG) under contracts WA 1116/28-1 and INST 184/154-1 for the X-ray diffractometer are gratefully acknowledged.

Conflicts of Interest: The authors declare no conflict of interest.

References

1. Chen, X.; Shen, S.; Guo, L.; Mao, S.S. Semiconductor-based Photocatalytic Hydrogen Generation. *Chem. Rev.* **2010**, *110*, 6503–6570. [CrossRef] [PubMed]
2. Hoffmann, R.M.; Martin, T.S.; Choi, W.; Bahnemann, W.D. Environmental Applications of Semiconductor Photocatalysis. *Chem. Rev.* **1995**, *95*, 69–96. [CrossRef]
3. Hu, C.; Hu, X.; Guo, J.; Qu, J. Efficient Destruction of Pathogenic Bacteria with $NiO/SrBi_2O_4$ under Visible Light Irradiation. *Environ. Sci. Technol.* **2006**, *40*, 5508–5513. [CrossRef] [PubMed]
4. Ng, H.Y.; Lightkap, V.I.; Goodwin, K.; Matsumura, M.; Kamat, V.P. To What Extent Do Graphene Scaffolds Improve the Photovoltaic and Photocatalytic Response of TiO_2 Nanostructured Films? *J. Phys. Chem. Lett.* **2010**, *1*, 2222–2227. [CrossRef]
5. Sakthivel, S.; Kisch, H. Daylight Photocatalysis by Carbon-Modified Titanium Dioxide. *Angew. Chem. Int. Ed.* **2003**, *42*, 4908–4911. [CrossRef] [PubMed]
6. Tong, T.; Zhang, J.; Tian, B.; Chen, F.; He, D. Preparation of Fe^{3+} doped TiO_2 catalysts by controlled hydrolysis of titanium alkoxide and study on their photocatalytic activity for methyl orange degradation. *J. Hazard. Mater.* **2008**, *155*, 572–579. [CrossRef]
7. Xu, J.J.; Ao, H.Y.; Fu, G.D.; Yuan, W.S. Synthesis of Gd-doped TiO_2 nanoparticles under mild condition and their photocatalytic activity. *Colloid Surf. A* **2009**, *334*, 107–111. [CrossRef]
8. Dodd, A.; Mckinley, A.; Tsuzuki, T.; Sauners, M. Optical and photocatalytic properties of nanoparticulate $(TiO_2)_x(ZnO)_{1-x}$ powders. *J. Alloys Compd.* **2010**, *489*, L17–L21. [CrossRef]
9. Dai, K.; Peng, Y.T.; Ke, N.D.; Wei, Q.B. Photocatalytic hydrogen generation using a nanocomposite of multi-walled carbon nanotubes and TiO_2 nanoparticles under visible light irradiation. *Nanotechnology* **2009**, *20*, 125603. [CrossRef]
10. Chai, B.; Peng, Y.T.; Zeng, P.; Mao, J. Synthesis of floriated In_2S_3 decorated with TiO_2 nanoparticles for efficient photocatalytic hydrogen production under visible light. *J. Mater. Chem.* **2011**, *21*, 14587–14593. [CrossRef]
11. Laokiat, L.; Khemthong, P.; Grisdanurak, N.; Sreearunothai, P.; Pattanasiriwisawa, W.; Klysubun, W. Photocatalytic degradation of benzene, toluene, ethylbenzene, and xylene (BTEX) using transition metal-doped titanium dioxide immobilized on fiberglass cloth. *Korean J. Chem. Eng.* **2012**, *29*, 377–383. [CrossRef]
12. Taffa, H.D.; Dillert, R.; Ulpe, C.A.; Bauerfeind, L.C.K.; Bredow, T.; Bahnemann, W.D.; Wark, M. Photoelectrochemical and theoretical investigations of spinel type ferrites ($M_xFe_{3-x}O_4$) for water splitting: A mini-review. *J. Photon. Energy* **2016**, *7*, 12009. [CrossRef]
13. Su, H.M.; He, C.; Sharma, K.V.; Abou Asi, M.; Xia, D.; Li, Z.X.; Deng, Q.H.; Xiong, Y. Mesoporous zinc ferrite: Synthesis, characterization, and photocatalytic activity with H_2O_2/visible light. *J. Hazard. Mater.* **2012**, *211–212*, 95–103. [CrossRef]
14. Cao, W.S.; Zhu, J.Y.; Cheng, F.G.; Huang, H.Y. $ZnFe_2O_4$ nanoparticles: Microwave-hydrothermal ionic liquid synthesis and photocatalytic property over phenol. *J. Hazard. Mater.* **2009**, *171*, 431–435. [CrossRef]
15. Casbeer, E.; Sharma, K.V.; Li, Z.X. Synthesis and photocatalytic activity of ferrites under visible light. A review. *Sep. Purif. Technol.* **2012**, *87*, 1–14. [CrossRef]
16. Nakanishi, T.; Masuda, Y.; Koumoto, K. Site-Selective Deposition of Magnetite Particulate Thin Films on Patterned Self-assembled Monolayers. *Chem. Mater.* **2004**, *16*, 3484–3488. [CrossRef]
17. Juan, X.W.; Yun, H.S.; Yan, H.T.; Hua, Q.Y. Photocatalytic Degradation of Water-Soluble Azo Dyes by $LaFeO_3$ and $YFeO_3$. *Adv. Mater. Res.* **2012**, *465*, 37–43.

18. Hou, L.; Sun, G.; Liu, K.; Li, Y.; Gao, F. Preparation, characterization and investigation of catalytic activity of Li-doped LaFeO$_3$ nanoparticles. *J. Sol-Gel Sci. Technol.* **2006**, *40*, 9–14. [CrossRef]
19. Tang, P.; Fu, M.; Chen, H.; Cao, F. Synthesis of Nanocrystalline LaFeO$_3$ by Precipitation and its Visible-Light Photocatalytic Activity. *Mater. Sci. Forum* **2011**, *694*, 150–154. [CrossRef]
20. Thirumalairajan, S.; Girija, K.; Masteralo, R.V.; Ponpandian, N. Photocatalytic degradation of organic dyes under visible light irradiation by floral-like LaFeO$_3$ nanostructures comprised of nanosheet petals. *New J. Chem.* **2014**, *38*, 5480–5490. [CrossRef]
21. Su, H.; Jing, L.; Shi, K.; Yao, C.; Fu, H. Synthesis of large surface area LaFeO$_3$ nanoparticles by SBA-16 template method as high active visible photocatalysts. *J. Nanopart. Res.* **2010**, *12*, 967–974. [CrossRef]
22. Yang, J.; Zhong, H.; Li, M.; Zhang, L.; Zhang, Y. Markedly enhancing the visible-light photocatalytic activity of LaFeO$_3$ by post-treatment in molten salt. *React. Kinet. Catal. Lett.* **2009**, *97*, 269–274. [CrossRef]
23. Tijare, N.S.; Joshi, V.M.; Padole, S.P.; Mangukar, A.P.; Rayalu, S.S.; Labhsetwar, K.N. Photocatalytic hydrogen generation through water splitting on nano-crystalline LaFeO$_3$ perovskite. *Int. J. Hydrog. Energy* **2012**, *37*, 10451–10456. [CrossRef]
24. Wu, H.; Hu, R.; Zhou, T.; Li, C.; Meng, W.; Yang, J. A novel efficient boron-doped LaFeO$_3$ photocatalyst with large specific surface area for phenol degradation under simulated sunlight. *CrystEngComm* **2015**, *17*, 3859–3865. [CrossRef]
25. Ju, L.; Chen, Z.; Fang, L.; Dong, W.; Zheng, F.; Shen, M. Sol-Gel Synthesis and Photo-Fenton-Like Catalytic Activity of EuFeO$_3$ Nanoparticles. *J. Am. Ceram. Soc.* **2011**, *94*, 3418–3424. [CrossRef]
26. Thirumalairajan, S.; Girija, K.; Ganesh, I.; Mangalaraj, D.; Viswanathan, C.; Balamurugan, A. Controlled synthesis of perovskite LaFeO$_3$ microsphere composed of nanoparticles via self-assembly process and their associated photocatalytic activity. *Chem. Eng. J.* **2012**, *209*, 420–428. [CrossRef]
27. Dong, G.; Zhang, L. Porous structure dependent photoreactivity of graphitic carbon nitride under visible light. *J. Mater. Chem.* **2012**, *22*, 1160–1166. [CrossRef]
28. Li, K.; Wang, D.; Wu, F.; Xie, T.; Li, T. Surface electronic states and photovoltage gas-sensitive characters of nanocrystalline LaFeO$_3$. *Mater. Chem. Phys.* **2000**, *64*, 269–272. [CrossRef]
29. Marschall, R.; Soldat, J.; Wark, M. Enhanced photocatalytic hydrogen generation from barium tantalate composites. *Photochem. Photobiol. Sci.* **2013**, *12*, 671–677. [CrossRef] [PubMed]
30. Qi, X.; Zhou, J.; Yue, Z.; Gui, Z.; Li, L. Auto-combustion synthesis of nanocrystalline LaFeO$_3$. *Mater. Chem. Phys.* **2002**, *78*, 25–29. [CrossRef]
31. Rida, K.; Benabbas, A.; Bouremmad, F.; Pena, A.M.; Sastre, E.; Martinez, A. Effect of calcination temperature on the structural characteristics and catalytic activity for propene combustion of sol-gel derived lanthanum chromite perovskite. *Appl. Catal. A* **2007**, *327*, 173–179. [CrossRef]
32. Thirumalairajan, S.; Girija, K.; Hebalkar, Y.N.; Mangalaraj, D.; Viswanathan, C.; Ponpandian, N. Shape evolution of perovskite LaFeO$_3$ nanostructures: A systematic investigation of growth mechanism, properties and morphology dependent photocatalytic activity. *RSC Adv.* **2013**, *3*, 7549–7561. [CrossRef]
33. Zhang, J.; Li, M.; Feng, Z.; Chen, J.; Li, C. UV Raman spectroscopic study on TiO$_2$. I. Phase transformation at the surface and in the bulk. *J. Phys. Chem. B* **2006**, *110*, 927–935. [CrossRef]
34. Wu, W.; Liang, S.; Shen, L.; Ding, Z.; Zheng, H.; Su, W.; Wu, L. Preparation, characterization and enhanced visible light photocatalytic activities of polyaniline/Bi$_3$NbO$_7$ nanocomposites. *J. Alloys Compd.* **2012**, *520*, 213–219. [CrossRef]
35. Yu, L.; Zhang, X.; Li, G.; Cao, Y.; Shao, Y.; Li, D. Highly efficient Bi$_2$O$_2$CO$_3$/BiOCl photocatalyst based on heterojunction with enhanced dye-sensitization under visible light. *Appl. Catal. B Environ.* **2016**, *187*, 301–309. [CrossRef]
36. Ismael, M.; Elhaddad, E.; Taffa, H.D.; Wark, M. Synthesis of Phase Pure Hexagonal YFeO$_3$ Perovskite as Efficient Visible Light Active Photocatalyst. *Catalysts* **2017**, *7*, 326. [CrossRef]
37. Parida, M.K.; Reddy, H.K.; Martha, S.; Das, P.D.; Biswal, N. Fabrication of nanocrystalline LaFeO$_3$: An efficient sol-gel auto-combustion assisted visible light responsive photocatalyst for water decomposition. *Int. J. Hydrog. Energy* **2010**, *35*, 12161–12168. [CrossRef]
38. Vaiano, V.; Iervolino, G.; Sannino, D. Enhanced Photocatalytic Hydrogen Production from Glucose on Rh-Doped LaFeO$_3$. *Chem. Eng. Trans.* **2017**, *60*, 235–240.
39. Xu, K.; Feng, J. Superior photocatalytic performance of LaFeO$_3$/gC$_3$N$_4$ heterojunction nanocomposites under visible light irradiation. *RSC Adv.* **2017**, *7*, 45369–45376. [CrossRef]

40. Yang, H.; Yan, J.; Lu, Z.; Cheng, X.; Tang, Y. Photocatalytic activity evaluation of tetragonal CuFe$_2$O$_4$ nanoparticles for the H$_2$ evolution under visible light irradiation. *J. Alloys Compd.* **2009**, *476*, 715–719. [CrossRef]
41. Rekhila, G.; Bessekhouad, Y.; Trari, M. Visible light hydrogen production on the novel ferrite NiFe$_2$O$_4$. *Int. J. Hydrog. Energy* **2013**, *38*, 6335–6343. [CrossRef]
42. Chen, X.; Zhou, B.; Yang, S.; Wu, H.; Wu, Y.; Wu, Y.; Pan, J.; Xiong, X. In situ construction of a SnO$_2$/g-C$_3$N$_4$ heterojunction for enhanced visible-light photocatalytic activity. *RSC Adv.* **2015**, *5*, 68953–68963. [CrossRef]
43. He, M.Y.; Cai, J.; Zhang, H.L.; Wang, X.X.; Lin, J.H.; Teng, T.B.; Zhao, H.L.; Weng, Z.W.; Wan, L.H.; Fan, M. Comparing Two New Composite Photocatalysts, t-LaVO$_4$/g-C$_3$N$_4$ and m-LaVO$_4$/g-C$_3$N$_4$, for Their Structures and Performances. *Ind. Eng. Chem. Res.* **2014**, *53*, 5905–5915. [CrossRef]
44. Yu, T.H.; Quan, X.; Chen, S.; Zhao, M.H.; Zhang, B.Y. TiO$_2$–carbon nanotube heterojunction arrays with a controllable thickness of TiO$_2$ layer and their first application in photocatalysis. *J. Photochem. Photobiol. A Chem.* **2008**, *200*, 301–306. [CrossRef]
45. Lim, J.; Monllor-Satocaa, D.; Jang, S.J.; Lee, S.; Choi, W. Visible light photocatalysis of fullerol-complexed TiO$_2$ enhanced by Nb doping. *Appl. Catal. B Environ.* **2014**, *152–153*, 233–240. [CrossRef]
46. Bi, P.Y.; Quyang, X.S.; Cao, Y.J.; Ye, H.J. Facile synthesis of rhombic dodecahedral AgX/Ag$_3$PO$_4$ (X = Cl, Br, I) heterocrystals with enhanced photocatalytic properties and stabilities. *Phys. Chem. Chem. Phys.* **2011**, *13*, 10071–10075. [CrossRef]
47. Hong, S.; Lee, S.; Jang, S.J.; Lee, S.J. Heterojunction BiVO$_4$/WO$_3$ electrodes for enhanced photoactivity of water oxidation. *Energy Environ. Sci.* **2011**, *4*, 1781–1787. [CrossRef]
48. Li, X.; Ye, J. Photocatalytic Degradation of Rhodamine B over Pb$_3$Nb$_4$O$_{13}$/Fumed SiO$_2$ Composite under Visible Light Irradiation. *J. Phys. Chem. C* **2007**, *111*, 13109–13116. [CrossRef]
49. Merka, O.; Yarovyi, V.; Bahnemann, D.W.; Wark, M. pH-Control of the Photocatalytic Degradation Mechanism of Rhodamine B over Pb$_3$Nb$_4$O$_{13}$. *J. Phys. Chem. C* **2011**, *115*, 8014–8023. [CrossRef]
50. Wang, X.; Zhang, L.; Lin, H.; Nong, Q.; Wu, Y.; Wu, T.; He, Y. Synthesis and characterization of ZrO$_2$/g-C$_3$N$_4$ composite with enhanced visible-light photoactivity for rhodamine degradation. *RSC Adv.* **2014**, *4*, 40029–40035. [CrossRef]
51. Ohtani, B. Photocatalysis A to Z- what we know and what we do not know in a scientific sense. *J. Photochem. Photobiol. C Photochem. Rev.* **2010**, *11*, 157–178. [CrossRef]
52. Liu, W.; Wang, M.; Xu, C.; Chen, S. Chem. Facile synthesis of g-C$_3$N$_4$/ZnO composite with enhanced visible light photooxidation and photoreduction properties. *Chem. Eng. J.* **2012**, *209*, 386–393. [CrossRef]
53. Yu, J.; Dai, G.; Cheng, B. Effect of Crystallization Methods on Morphology and Photocatalytic Activity of Anodized TiO$_2$ Nanotube Array Films. *J. Phys. Chem. C* **2010**, *114*, 19378–19385. [CrossRef]
54. He, M.Y.; Cai, J.; Li, T.T.; Wu, Y.; Lin, J.H.; Zhao, H.L.; Luo, F.M. Efficient degradation of RhB over GdVO$_4$/g-C$_3$N$_4$ composites under visible-light irradiation. *Chem. Eng. J.* **2013**, *215–216*, 721–730. [CrossRef]
55. Yang, X.; Cui, H.; Li, Y.; Qin, J.; Zhang, R.; Tang, H. Fabrication of Ag$_3$PO$_4$-Graphene Composites with Highly Efficient and Stable Visible Light Photocatalytic Performance. *ACS Catal.* **2013**, *3*, 363–369. [CrossRef]
56. Wang, F.D.; Kako, T.; Ye, H.J. Efficient Photocatalytic Decomposition of Acetaldehyde over a Solid-Solution Perovskite (Ag$_{0.75}$Sr$_{0.25}$)(Nb$_{0.75}$Ti$_{0.25}$)O$_3$ under Visible-Light Irradiation. *J. Am. Chem. Soc.* **2008**, *130*, 2724–2725. [CrossRef]
57. Kormali, P.; Triantis, T.; Dimotikali, D.; Hiskia, A.; Papaconstantinou, E. On the photooxidative behavior of TiO$_2$ and PW$_{12}$O$_{40}$$^{3-}$: OH radicals versus holes. *Appl. Catal. B* **2006**, *68*, 139–146. [CrossRef]
58. Chen, Y.-L.; Zhang, D.-W. A Simple Strategy for the preparation of g-C$_3$N$_4$/SnO$_2$ Nanocomposite Photocatalysts. *Sci. Adv. Mater.* **2014**, *6*, 1091–1098. [CrossRef]
59. Pirzada, M.B.; Kunchala, K.R.P.; Naidu, S.B. Synthesis of LaFeO$_3$/Ag$_2$CO$_3$ Nanocomposites for Photocatalytic Degradation of Rhodamine B and p-Chlorophenol under Natural Sunlight. *ACS Omega* **2019**, *4*, 2618–2629. [CrossRef]
60. Hu, R.; Li, C.; Wang, X.; Sun, Y.; Jia, H.; Su, H.; Zhang, Y. Photocatalytic activities of LaFeO$_3$ and La$_2$FeTiO$_6$ in p-chlorophenol degradation under visible light. *Catal. Commun.* **2012**, *29*, 35–39. [CrossRef]

© 2019 by the authors. Licensee MDPI, Basel, Switzerland. This article is an open access article distributed under the terms and conditions of the Creative Commons Attribution (CC BY) license (http://creativecommons.org/licenses/by/4.0/).

Article

On the Effects of Doping on the Catalytic Performance of (La,Sr)CoO$_3$. A DFT Study of CO Oxidation

Antonella Glisenti [1,2] and Andrea Vittadini [1,2,*]

1. Dipartimento di Scienze Chimiche, Università di Padova, via Marzolo 1, I-35131 Padova PD, Italy; antonella.glisenti@unipd.it
2. CNR-ICMATE, via Marzolo 1, I-35131 Padova PD, Italy
* Correspondence: andrea.vittadini@unipd.it; Tel.: +39-049-827-5235

Received: 26 February 2019; Accepted: 25 March 2019; Published: 30 March 2019

Abstract: The effects of modifying the composition of LaCoO$_3$ on the catalytic activity are predicted by density functional calculations. Partially replacing La by Sr ions has beneficial effects, causing a lowering of the formation energy of O vacancies. In contrast to that, doping at the Co site is less effective, as only 3d impurities heavier than Co are able to stabilize vacancies at high concentrations. The comparison of the energy profiles for CO oxidation of undoped and of Ni-, Cu-m and Zn-doped (La,Sr)CoO$_3$(100) surface shows that Cu is most effective. However, the effects are less spectacular than in the SrTiO$_3$ case, due to the different energetics for the formation of oxygen vacancies in the two hosts.

Keywords: heterogeneous catalysis; surface science; materials science; perovskites; CO oxidation; DFT calculations; transition metal doping

1. Introduction

As internal combustion engines are likely to power most world vehicles for at least the next two decades, while legislations are tightening the emission limits, the reduction of noxious components in the exhaust gas stands as a priority for the automotive industry. The main tool for achieving this result is provided by three way catalysts (TWCs) installed downstream of engines. Unfortunately, state-of-the-art TWCs typically contain platinum group metals and/or other elements such as rare earths, whose demand is rapidly rising, and whose production occurs in unevenly distributed areas. Hence, finding new and sustainable materials for TWCs is a current issue for research in catalysis. In this regard, a convenient and well known approach to tune the electronic and, consequently, the catalytic properties of materials, is doping with transition metal atoms [1]. In fact, experimental and theoretical investigations [2–4] have shown that even inert compounds, such as SrTiO$_3$, can be turned into effective catalysts for CO oxidation if doped with transition metals, such as Co or Cu. Furthermore, it has been established that the primary role of doping it that of enhancing the surface oxygen atoms activity, which is gauged by the formation energy of surface oxygen vacancies. In fact, the rate determining step of the CO oxidation reaction is the extraction of surface oxygen by CO [3]. On the other hand, a more active host, such as LaCoO$_3$, can improve its performance as a TWC material by partially replacing Co with Cu [5]. This is, however, most effective for high doping levels, when copper oxides segregate at the surface. Hence, it is not clear whether a low concentration of Cu dopant can actually improve the activity of LaCoO$_3$ for TWC applications. Furthermore, we are not aware of systematic studies comparing the effects of different transition metals as dopants on the catalytic properties of LaCoO$_3$. The aim of the present work is to understand whether doping can improve the performance of LaCoO$_3$ as a material for three-way catalytic converters. In particular, we want to assess whether the vacancy formation energy is a good a descriptor for the catalytic properties of perovskites, as it was found in our recent studies on the protypical $SrTiO_3$ perovskite.

2. Results

2.1. $LaCoO_3$ vs. $(La,Sr)CoO_3$

As the stoichiometry of perovskites can be changed by replacing both the A-site and the B-site ions, the range of possible stoichiometries is quite large. Thus, we decided to adopt a simple approach where we first compare the properties of the $LaCoO_3$ (LCO) and of $(La,Sr)CoO_3$ (LSCO) hosts. For the latter, we assumed a $La_{0.75}Sr_{0.25}CoO_3$ stoichiometry, which has been computed to be a stable phase by Fuks et al. [6]. We considered the formation of both single and double oxygen vacancies (V_O), as well as the adsorption of CO and NO molecules at the CoO-terminated (100) surface, testing all the configurations allowed by a 2 × 2 supercell. The main results are reported in Table 1.

Table 1. Formation energy for surface vacancies and CO adsorption energies (eV) at the (100) surface of LCO and LSCO.

Compound	1 V_O/cell	2 V_O/cell	CO Adsorption	NO Adsorption
LCO	2.12	5.15	0.98	1.10
LSCO	1.83	4.37	0.90	1.16

Interestingly, the partial substitution of La by Sr atoms enhances the stability of vacancies, while having minor (and contrasting) effects on the adsorption of CO and NO. Hence, this replacement is in principle suitable for tuning the catalytic properties of $LaCoO_3$. Another interesting finding is that when two vacancies are introduced in the supercell, they prefer to cluster together.

2.2. Doping LSCO at the Co Site

On the basis of the above presented results, it seems interesting to investigate the effects of doping LSCO at the Co site, once more taking the vacancy formation energy as a gauge of the catalytic activity. To this end, we replace one of the surface Co atoms of the supercell by another 3d atom, namely V, Cr, Mn, Fe, Ni, Cu, and Zn. Next, we compute the stability of oxygen vacancies, taking into consideration both the sites adjacent (NN) and those not-adjacent (NNN) to the 3d-impurity. The same slab models are used to compute the adsorption energy of CO, which is in turn evaluated both at the impurity site (CO@M) and at regular cobalt sites (CO@Co). We resume the results of these calculations in Figure 1, from which the difference in the behaviour of the investigated systems can be inferred. In particular, it appears that though the formation of vacancies is favored both by light and by heavy 3d metal impurities (with the exception of Fe), vacancies are attracted only by light dopants. Furthermore, CO adsorption is always favored at regular (unsubstituted) Co sites.

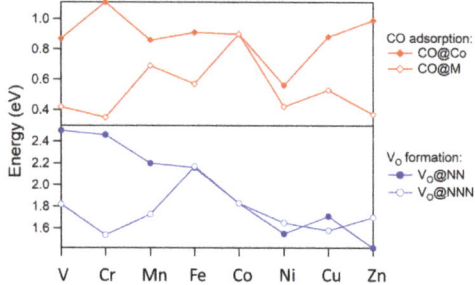

Figure 1. CO adsorption energies (top) and vacancy formation energies (bottom) computed for TM-doped LSCO(100) surfaces. The CO@M and CO@Co lines indicate adsorption at impurity and at regular Co sites, respectively. Vacancies created at the nearest site and at the next nearest site of the impurity are labelled NN and NNN, respectively.

We now consider the effect of Co-site doping on the formation of double vacancies. This is done starting from the structure of single vacancies, and exploring the stability of the possible configurations corresponding to the creation of a second vacancy. The results, shown in Figure 2, indicate that, in contrast to the case of single vacancies, the formation of double vacancies is favored only by heavy 3d impurities, namely, Ni, Cu, and Zn. In contrast to that, doping with light 3d metals is detrimental, as it increases the formation energy of double vacancies by ~0.5 eV wrt the undoped LSCO host (corresponding to the Co label in Figure 2) which in turn indicates a worse performance for NO reduction. Therefore, doping with light 3d impurities, such as V, Cr, Mn, and Fe, is not a wise choice for improving the properties of LSCO in TWC converters.

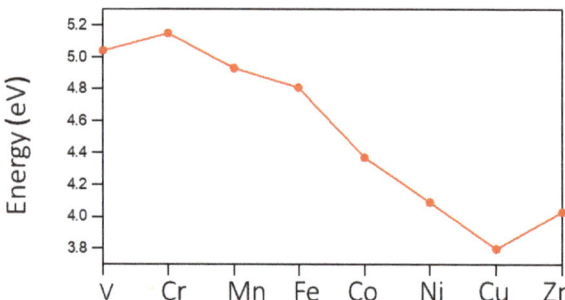

Figure 2. Formation energy for double vacancies (i.e., 2 V_O/cell) for LSCO doped at the Co site.

2.3. CO Oxidation at LSCO: Effects of Doping at the Co Site

The results presented in the preceding section encourage us to investigate in more detail how doping LSCO with heavy 3d dopants, i.e. Ni, Cu, and Zn, can influence the energetics of the CO oxidation reaction. To this end, we consider the mechanism reported in Ref. [3], consisting of the following elementary steps: i. Adsorption of CO; ii. Abstraction of lattice O atom and formation of CO_2; iii. Desorption of CO_2; iv. Adsorption of O_2; v. Capture of a second CO molecule and formation of CO_2; and vi. Desorption of CO_2 and restoration of the stoichiometry of the starting surface. On the basis of the above presented findings, CO has been assumed to be adsorbed at a regular Co site, and to abstract a nearby O oxygen to form CO_2. We optimized all the intermediate species and labelled them with consecutive numbers, starting from the clean surface (**1**). The optimized structures of these species turn out to be quite similar for the pure and for the doped systems, and are sketched in Figure 3 for the case of the Cu-doped surface. Actually, the structures are also similar to those of the analogous intermediates computed for SrTiO$_3$ [3] in the case of Cu doping. In particular, both O$_2$ and CO$_2$ lay parallel to the surface.

The analysis of the energy profiles relative to the investigated systems, reported in Figure 4, reveals that, similarly to what has been observed for the structures of the intermediates, the reaction energetics is also scarcely perturbed by doping, which is in striking contrast to the SrTiO$_3$ case [3]. In fact, step ii, where a surface O atom is abstracted by CO, is always exothermic by 1.2–1.3 eV for all the systems, with the exception of the Cu-doped sytem, where it is even more exothermic (~1.5 eV). Overall, the energy profile of the reaction is qualitatively very similar to that computed for Cu-doped SrTiO$_3$: all the steps are exothermic, with the exception of those corresponding to the desorption of weakly adsorbed molecules, which are slightly endothermic

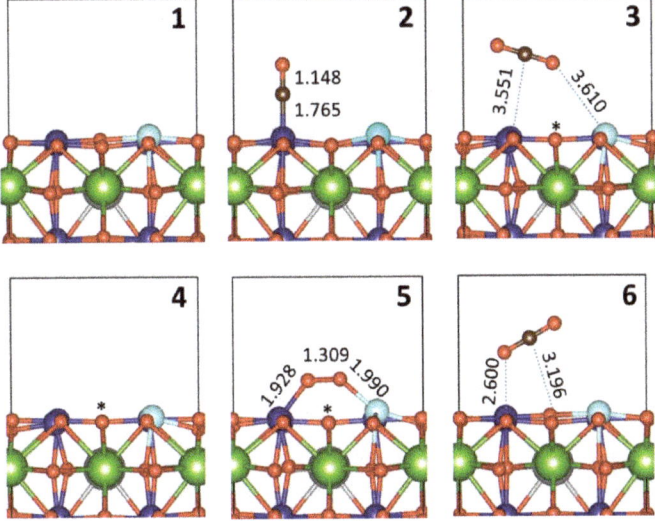

Figure 3. Intermediate species in the CO oxidation process on the Cu-doped LSCO(100) surface. The color codes are red = O, blue = Co, green = La, gray = Sr, cyan = Cu, brown = C. Asterisks mark the position of O vacancies.

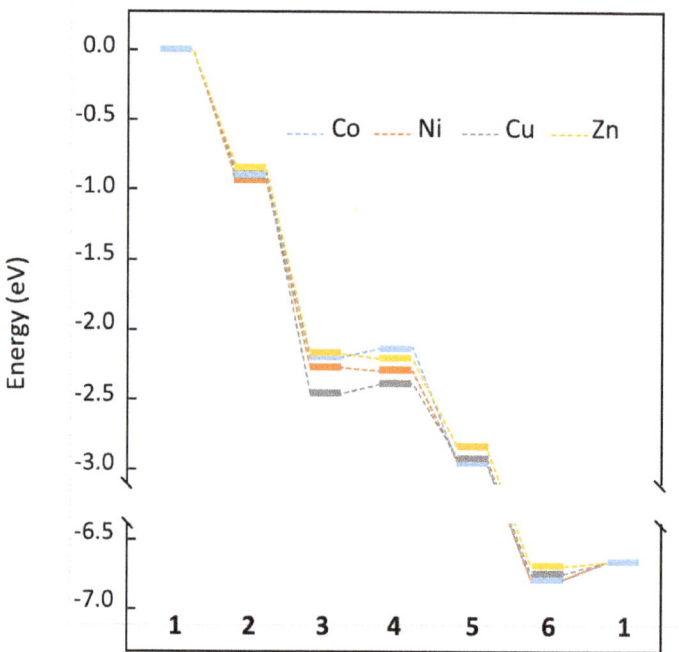

Figure 4. Energy profiles reporting all the intermediates (see Figure 3) in the CO oxidation reaction occurring on the pure as well as on the Ni-, Cu-, and Zn-doped LSCO(100) surfaces.

We now want to compare the potential energy (PE) curves of the O abstraction process (step ii, *viz.* 2 → 3) for the undoped and for the Cu-doped LSCO surfaces as obtained from nudged elastic band (NEB) calculations. In tune with the analogies found both in the structure of the intermediates, and in the energy profiles (see Figure 4), the PE curves, shown in Figure 5, appear to be quite similar, the barrier being slightly lower for the Cu-doped case (0.35 vs. 0.45 eV). Because of the exothermicity of the step, we find "earlier" transition states, with respect to the case of s $SrTiO_3$. In fact, transition state (TS) structures (see Figure 5) are similar to semi-bridging CO molecules, whereas CO_2-like structures were computed in the latter case [3]. For analogous reasons, the C-Co distance of the TS becomes shorter (1.811 vs. 1.765 Å) on passing from the undoped to the Cu-doped surface, while a slight shortening is computed also for the C-O distance (1.159 vs. 1.156 Å).

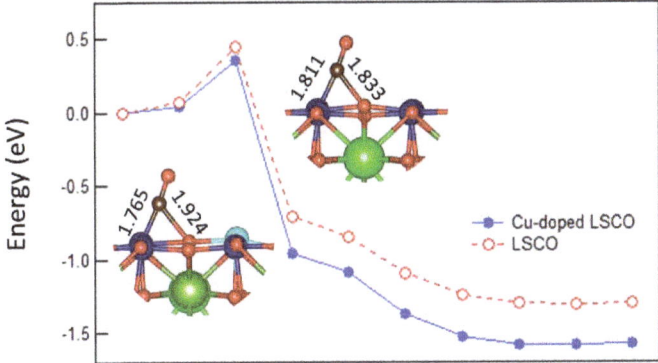

Figure 5. Potential energy surfaces for the oxygen abstraction step (**2 → 3**) on the pure and on the Cu-doped LSCO(100) surfaces. Inset pictures show the structure of the transition states. The colors are the same as in Figure 3.

3. Theoretical Methods

Similarly to our previous investigations of $SrTiO_3$, we used the PWSCF code of the QUANTUM-ESPRESSO package [7] to solve the spin-polarized Kohn-Sham equations with the generalized gradient approximation (GGA), adopting the PBE exchange-correlation functional [8]. We point out that our aim is not giving accurate descriptions of the electronic structure, nor quantitative predictions of any kind. Hence, we rely on standard GGA calculations, which have been successfully tested in similar cases [9,10], and are certainly affordable when comparing trends, instead of performing GGA+U calculations, which are time-consuming and often problematic to converge. The valence-core interaction were described by ultrasoft pseudopotentials taken from the Garrity–Bennett–Rabe–Vanderbilt library [11]. All the calculations have been initialized in the ferromagnetic state. Though most of the calculations have been carried out using the recommended 40 ryd plane wave kinetic energy cutoff, for transition state calculations (vide infra) a 30 ryd cutoff was used. We actually found that reducing the cutoff from 40 to 30 ryd has minor effects on the geometries, and produces a general 0.02–0.05 eV reduction of the adsorption energies. Lattice constants of $LaCoO_3$ and of $La_{0.75}Sr_{0.25}CoO_3$ have been optimized and kept fixed for all the subsequent calculations, as dopants have been treated as diluted impurities. Bulk structures have been studied with cubic supercells containing 40 atoms. For LSCO, Sr ions were assumed to occupy the lattice positions as in the structure proposed by Fuks et al. [6], i.e., they were placed at the farthest possible distance in a 2 × 2 × 2 supercell. We have considered the (100) surfaces, which have been modelled with a 2 × 2 slab consisting of seven atomic layers of CoO_2 and LaO stacked alternately. Only the top (CoO_2-terminated) surface of the slab was used to model adsorption and reactions. The top three atomic layers are relaxed, whereas the bottom four atomic layers were kept fixed to simulate bulk. Effects of impurities and oxygen vacancies were evaluated by modifying the composition of the top surface. Thanks to the symmetric termination, a 12 Å thick vacuum space is sufficient to decouple the surfaces. Increasing the vacuum thickness to 18 Å changes adsorption energies by less than 0.01 eV. The surface Brillouin zone was sampled using a 2 × 2 k-point mesh. Transition states (TSs) were located using the climbing image nudged elastic band (CI-NEB). Zero-point energy and entropic contributions were not included, as they cannot change the computed trends. In computing the formation energy of oxygen vacancies, a correction was applied to compensate the well-known DFT-GGA tendency to overestimate the O_2 dissociation energy [12].

4. Conclusions

We have investigated the effects of modifying the composition of LaCoO$_3$ using DFT calculations. The vacancy formation energy, taken as a gauge of the catalytic activity, is reduced both by introducing Sr atoms at the La site and by doping with 3d atoms at the Co site. Only heavy 3d atoms are able to promote high density of vacancies, which is needed for NO reduction. Overall, Cu-doped and Ni-doped (La,Sr)CoO$_3$ appear to be the best materials for preparation of catalysts suitable for three-way converters. It should, however, be pointed out that, due to the early nature of the transition state for the vacancy formation step in LSCO, the increased stability of oxygen vacancies obtained by doping is predicted to have a reduced influence on the catalytic properties when compared to that found for SrTiO$_3$. Clearly, this work represents a first step towards a full understanding of the catalytic properties of these complex systems. More efforts, both on the theoretical and on the experimental side, is needed to extend and corroborate our conclusions.

Author Contributions: Conceptualization, A.V.; methodology, A.V.; software, A.V.; validation, A.V..; formal analysis, A.V.; investigation, A.V.; resources, A.V.; data curation, A.V.; writing—original draft preparation, A.V.; writing—review and editing, A.G.; visualization, A.V.; supervision, A.V.; project administration, A.G.; funding acquisition, A.G.

Funding: This project has received funding from the European Union's Horizon 2020 research and innovation programme under grant agreement No 686086. This publication reflects only the author's view and the Commission is not responsible for any use that may be made of the information it contains.

Conflicts of Interest: The authors declare no conflict of interest.

References

1. McFarland, E.W.; Metiu, H. Catalysis by Doped Oxides. *Chem. Rev.* **2013**, *113*, 4391–4427. [CrossRef]
2. Carlotto, S.; Natile, M.M.; Glisenti, A.; Vittadini, A. Catalytic Mechanisms of NO Reduction in a CO-NO Atmosphere at Co- and Cu-Doped SrTiO$_3$(100) Surfaces. *J. Phys. Chem. C* **2018**, *122*, 449–454. [CrossRef]
3. Carlotto, S.; Natile, M.M.; Glisenti, A.; Paul, J.-F.; Blanck, D.; Vittadini, A. Energetics of CO oxidation on lanthanide-free perovskite systems: The case of Co-doped SrTiO$_3$. *Phys. Chem. Chem. Phys.* **2016**, *18*, 33282–33286. [CrossRef] [PubMed]
4. Glisenti, A.; Natile, M.M.; Carlotto, S.; Vittadini, A. Co- and Cu-Doped Titanates: Toward a New Generation of Catalytic Converters. *Catal. Lett.* **2014**, *144*, 1466–1471. [CrossRef]
5. Glisenti, A.; Pacella, M.; Guiotto, M.; Natile, M.M.; Canu, P. Largely Cu-doped LaCo$_{1-x}$Cu$_x$O$_3$ perovskites for TWC: Toward new PGM-free catalysts. *Appl. Catal. B Environ.* **2016**, *180*, 94–105. [CrossRef]
6. Fuks, D.; Weizman, A.; Kotomin, E. Phase competition in (La1-c,Sr-c)CoO$_3$ solid solutions: Ab initio thermodynamic study. *Phys. Status Solidi B Basic Solid State Phys.* **2013**, *250*, 864–869. [CrossRef]
7. Giannozzi, P.; Baroni, S.; Bonini, N.; Calandra, M.; Car, R.; Cavazzoni, C.; Ceresoli, D.; Chiarotti, G.L.; Cococcioni, M.; Dabo, I.; et al. QUANTUM ESPRESSO: A modular and open-source software project for quantum simulations of materials. *J. Phys. Condens. Matter* **2009**, *21*, 395502. [CrossRef] [PubMed]
8. Perdew, J.; Burke, K.; Ernzerhof, M. Generalized gradient approximation made simple. *Phys. Rev. Lett.* **1996**, *77*, 3865–3868. [CrossRef] [PubMed]
9. Choi, S.O.; Penninger, M.; Kim, C.H.; Schneider, W.F.; Thompson, L.T. Experimental and Computational Investigation of Effect of Sr on NO Oxidation and Oxygen Exchange for La$_{1-x}$Sr$_x$CoO$_3$ Perovskite Catalysts. *ACS Catal.* **2013**, *3*, 2719–2728. [CrossRef]
10. Penninger, M.W.; Kim, C.H.; Thompson, L.T.; Schneider, W.F. DFT Analysis of NO Oxidation Intermediates on Undoped and Doped LaCoO$_3$ Perovskite. *J. Phys. Chem. C* **2015**, *119*, 20488–20494. [CrossRef]
11. Garrity, K.F.; Bennett, J.W.; Rabe, K.M.; Vanderbilt, D. Pseudopotentials for high-throughput DFT calculations. *Comput. Mater. Sci.* **2014**, *81*, 446–452. [CrossRef]
12. Wang, L.; Maxisch, T.; Ceder, G. Oxidation energies of transition metal oxides within the GGA+U framework. *Phys. Rev. B* **2006**, *73*, 195107. [CrossRef]

 © 2019 by the authors. Licensee MDPI, Basel, Switzerland. This article is an open access article distributed under the terms and conditions of the Creative Commons Attribution (CC BY) license (http://creativecommons.org/licenses/by/4.0/).

Article

BaTi$_{0.8}$B$_{0.2}$O$_3$ (B = Mn, Fe, Co, Cu) LNT Catalysts: Effect of Partial Ti Substitution on NOx Storage Capacity

Craig Aldridge, Verónica Torregrosa-Rivero, Vicente Albaladejo-Fuentes, María-Salvadora Sánchez-Adsuar and María-José Illán-Gómez *

Departamento de Química Inorgánica, Facultad de Ciencias, Universidad de Alicante, Ap. 99, E-03080 Alicante, Spain; craigald@hotmail.co.uk (C.A.); vero.torregrosa@ua.es (V.T.-R.); vicentealbaladejo@gmail.com (V.A.-F.); dori@ua.es (M.-S.S.-A.)
* Correspondence: illan@ua.es; Tel.: +34-965903975

Received: 6 March 2019; Accepted: 16 April 2019; Published: 18 April 2019

Abstract: The effect of partial Ti substitution by Mn, Fe, Co, or Cu on the NOx storage capacity (NSC) of a BaTi$_{0.8}$B$_{0.2}$O$_3$ lean NOx trap (LNT) catalyst has been analyzed. The BaTi$_{0.8}$B$_{0.2}$O$_3$ catalysts were prepared using the Pechini's sol–gel method for aqueous media. The characterization of the catalysts (BET, ICP-OES, XRD and XPS) reveals that: i) the partial substitution of Ti by Mn, Co, or Fe changes the perovskite structure from tetragonal to cubic, whilst Cu distorts the raw tetragonal structure and promotes the segregation of Ba$_2$TiO$_4$ (which is an active phase for NOx storage) as a minority phase and ii) the amount of oxygen vacancies increases after partial Ti substitution, with the BaTi$_{0.8}$Cu$_{0.2}$O$_3$ catalyst featuring the largest amount. The BaTi$_{0.8}$Cu$_{0.2}$O$_3$ catalyst shows the highest NSC at 400 °C, based on NOx storage cyclic tests, which is within the range of highly active noble metal-based catalysts.

Keywords: perovskite; NO to NO$_2$ oxidation; NOx storage capacity; LNT catalysts

1. Introduction

Diesel engines are a type of lean burn engine, operating at upper stoichiometric air-to-fuel ratios (A/F>14.7/1), which grew in popularity at the end of 20th century as they offered higher fuel efficiency and less CO$_2$ emissions respect to gasoline engines [1]. However, these engines show a highly relevant drawback since they generate large amounts of NOx and soot [2,3]. In order to minimize the level of these pollutants, more stringent standards were progressively established all over the world. Nowadays, it is accepted that the current EURO VI standard regarding NOx emissions is not met by just improving the quality of the fuel, by modifying the engine, or by using three-way-catalysts (TWCs). Consequently, alternative catalytic strategies are mandatory in order to avoid the disappearance of vehicles fitted with a diesel engine [4,5].

Two methodologies have been proposed to control NOx emission in lean burn engines: selective catalytic reduction (SCR) and NOx storage and reduction (NSR), also called lean NOx trapping (LNT). LNT technology involves the adsorption of NOx under lean conditions, followed by the periodic regeneration of the catalyst by reduction under rich conditions [6]. The conventional LNT catalysts (fitted in diesel cars) are composed of a platinum-group metal and an alkaline or alkaline-earth oxide (BaO or K$_2$O) supported on a high surface area material (Al$_2$O$_3$, TiO$_2$, …). It has been found that LNT technology matched with a TWC in a direct-injection spark ignition (DISI) engine can feature interesting results for NOx control emission [7].

Nevertheless, these conventional LNT catalysts present some drawbacks [8], with the high cost of noble metals (mainly Pt) being one of the most relevant. In fact, an interesting challenge has been highlighted in a recent EU report, regarding the need to develop alternatives to the use of critical raw materials such as precious metals [9]. In this line, the potential of perovskite base catalysts is being largely illustrated in the literature for environmental applications [10–12].

In previous studies [11,12], titanium was partially substituted by copper in the $BaTiO_3$ perovskite structure, showing the resulting $BaTi_{1-x}Cu_xO_3$ perovskites a high activity for NOx storage, which was attributed to the presence of oxygen vacancies (created on the catalyst surface as a consequence of the copper incorporation into the structure) and to the segregation of some phases (mainly $BaCO_3$ and Ba_2TiO_4, but also CuO). It was also concluded that the $BaTi_{0.8}Cu_{0.2}O_3$ catalyst presents a NOx storage capacity (NSC) at 420 °C in the range of levels reported for noble metal-based catalysts (around 300 µmol/g) [13], and hence could be proposed as a potential component of high-temperature LNT systems for lean burn engines, such as gasoline direct injection engines. Moreover, in the literature, other metals such as Mn, Fe, or Co have been proposed as promising B cations in the perovskite used as catalysts for NOx and soot removal [14–16]. Thus, the aim of this paper is to determine the effect of Ti partial substitution by Mn, Fe, and Co in the NSC of the $BaTi_{0.8}B_{0.2}O_3$ LNT catalyst. The results will be analyzed with respect to the performance of the previously studied $BaTi_{0.8}Cu_{0.2}O_3$ catalyst [11,12].

2. Results and Discussion

2.1. Characterization of the Catalysts

Table 1 presents the nomenclature and the basic characterization data of the catalysts: B metal content (measured by ICP-OES), BET surface area (obtained by applying the BET equation to N_2 adsorption data) and XPS. As it can be observed, the ICP-OES results reveal that all the metals added during the synthesis processes are present in the catalysts. Besides, the BET surface areas of the catalysts are low (as correspond to solids with negligible porosity, as mixed oxides with perovskite structure are [11]) and they range from 5 to 13 m^2/g.

Table 1. Nomenclature and basic characterization data.

Catalyst	Nomenclature	S_{BET} (m^2/g)	Bexp (wt%)/Bnom (wt%)	B/Ba+Ti+B [1]	$O_{lattice}$/Ba+Ti+B [2]
$BaTi_{0.8}Mn_{0.2}O_3$	BTMnO_2	13	5.2/5.4	0.08	1.5
$BaTi_{0.8}Fe_{0.2}O_3$	BTFeO_2	7	4.7/4.8	0.09	1.8
$BaTi_{0.8}Co_{0.2}O_3$	BTCoO_2	5	4.9/4.9	0.13	1.4
$BaTi_{0.8}Cu_{0.2}O_3$	BTCuO_2	12	4.9/5.0	0.07	1.4
$BaTiO_3$	BTO_ref	9	–	–	2.0

[1] B/Ba+Ti+B nominal = 0.1, [2] $O_{lattice}$/Ba+Ti+B nominal = 1.5.

Concerning the XRD results, Figure 1a shows the XRD patterns of the catalysts that reveal a perovskite like structure (the diffraction peaks observed at 2θ: 22.3°; 31.4°; 38.8°; 45.2°; 51.0°; 56.1°; 65.8°; 74.9°; for (100), (110), (111), (200), (210), (211), (220), and (310) lattice planes, correspond to the standard JCPDS for tetragonal perovskite structure: 5-626 [17]) as the major crystalline phase for all the catalysts. Based on the splitting of peak around 51°, it seems that the perovskite structure is tetragonal for BTO_ref and BTCuO_2, but it changes to cubic for the other catalysts [11,12]. The magnification of the main peak of the diffractograms (31.5°), included in Figure 1b, clearly shows a shift to a lower angle value, respect to the BTO reference which is more evident for BTMnO_2, BTCoO_2, and BTFeO_2 (suggesting a modification of the perovskite structure) [11,12] than for BTCuO_2 (as the tetragonal structure is preserved for this catalyst), even though a decrease in the peak intensity is featured. Additionally, other minority phases are also identified by XRD, that is, mainly Ba_2TiO_4 and $BaCO_3$ (formed by the carbonation of segregated barium oxide during samples atmospheric exposure) but also CuO for BTCuO_2, and only $BaCO_3$ for BTMnO_2, BTFeO_2, and BTCoO_2 catalysts. As it has been previously reported [11,12], the existence of segregated phases on a metal substituted perovskite proves that the metal has been incorporated into the perovskite structure. Therefore, the shift of the main diffraction peaks ascribed to the perovskite structure and the segregation of minority phases identified in the XRD patterns indicate that Ti is successfully substituted by Mn, Fe, Co, and Cu in the perovskite framework. It is worth mentioning that: i) for BTCoO_2 catalysts, other cobalt phases (BaCoO$_3$ perovskite and Co$_3$O$_4$) are also identified as minority segregated phases in the XRD diffractogram, suggesting that cobalt has been introduced into the perovskite lattice in a lower extent than Cu, Fe, and Mn; and ii) Ba$_2$TiO$_4$, which has been suggested as an active phase for NOx storage [11,12], is only detected for BTCuO_2 catalyst.

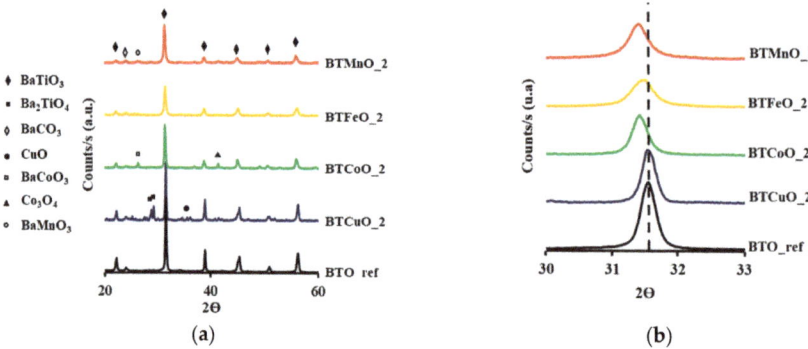

Figure 1. Catalysts characterization: (a) XRD patterns, (b) main peak magnification.

To verify the structural modifications in the perovskites suggested by XRD patterns, Raman spectroscopy was used. According to literature [18,19], only the tetragonal structure of BaTiO$_3$ perovskite, which belongs to space group P4mm, presents first-order Raman-active modes with bands at, approximately, 180 cm^{-1}, 265 cm^{-1}, 305 cm^{-1}, 520 cm^{-1}, and 720 cm^{-1}, corresponding to irreducible representations ((A1(LO)), (A1(TO)), (B1), (A1, E(TO)), and (A1, E(LO)), respectively. On the one hand, the Raman spectra, shown in Figure 2, confirm that BTCuO_2 preserves the original tetragonal structure of the raw perovskite, as suggested by XRD, as it features the main bands previously indicated. In spite of this, BTCuO_2 spectrum shows broader peaks than BTO spectrum, pointing out that copper incorporation distorts the original tetragonal structure. On the other hand, BTFeO_2 and BTMnO_2 catalysts show an almost flat spectrum, ascribed to perovskite cubic structure, which does not show active modes in Raman spectroscopy. Finally, some Raman peaks are identified in the BTCoO_2 spectrum which are ascribed to the presence of minority phases such as BaCoO$_3$ and Co$_3$O$_4$, also identified by XRD. This result supports that a lower degree of cobalt is incorporated into the perovskite framework of the catalyst.

The different effects of the B cations, that partially substitute Ti on the perovskite structure, seem to be related with their ionic radius. Fe, Mn, and Cu (as M^{2+}) have ionic radii larger than Ti^{4+} causing the distortion of the raw tetragonal perovskite structure. As Co^{2+} presents the most similar ionic radius to Ti^{4+}, the formation of the stable BaCoO$_3$ perovskite is also allowed and, consequently, a lower fraction of cobalt is inserted into the BaTiO$_3$ perovskite framework to partially replace titanium.

XPS provides valuable information about the catalysts surface composition. All the XPS spectra and contributions assignment are featured in Figure A1 and Table A1, in the Appendix A. Table 1 shows the data related to metal (Mn, Fe, Co, or Cu) distribution presented as B/Ba+Ti+B (B= Mn, Fe, Co, Cu) ratio, whilst the data related to lattice oxygen, is shown as O$_{lattice}$/Ba+Ti+B (B= Mn, Fe, Co, Cu) ratio. It can be observed that for Mn, Fe, and Cu, the B/Ba+Ti+B XPS ratio is lower than the corresponding nominal value (0.1), which supports that these metals have been partially introduced into the perovskite structure. The BTCuO_2 catalyst presents the lowest value, so, the highest percentage of metal inside the perovskite lattice, whilst for Co, a B/Ba+Ti+B ratio higher than the nominal is found due to the presence of BaCoO$_3$ and Co$_3$O$_4$ segregated phases. The O$_{lattice}$/Ba+Ti+B XPS ratio (calculated from the area for O1s peak corresponding to lattice oxygen) for all catalysts is lower than the corresponding value for the BTO_ref perovskite (2.0), evidencing the creation of oxygen vacancies in the perovskite structure to compensate the imbalance in positive charge due to the partial substitution of Ti^{4+}. Note that the BTCuO_2 catalyst presents the lowest O$_{lattice}$/Ba+Ti+B ratio and, consequently, the largest amount of surface oxygen vacancies. This result seems to be explained considering that BTCuO_2 catalyst presents a positive imbalance larger than the other catalysts due to the highest difference between the oxidation state of Ti^{4+} and the Cu^{+2} (Fe and Mn appear mainly as Fe(II) and Mn(III) but Fe(III) and Mn(IV) have been also identified by XPS). Additionally, surface oxygen vacancies are also created because Ti^{+4} cannot achieve a higher oxidation state as other B cations (as Mn or Fe in BaMn$_{1-x}$Cu$_x$O$_3$ and BaFe$_{1-x}$Cu$_x$O$_3$ [20,21]) do. Finally, it is remarkable that BTCuO_2 catalyst preserves the original tetragonal structure (shown by XRD and Raman results), but with a high degree of distortion, which causes the presence of a larger amount of oxygen vacancies.

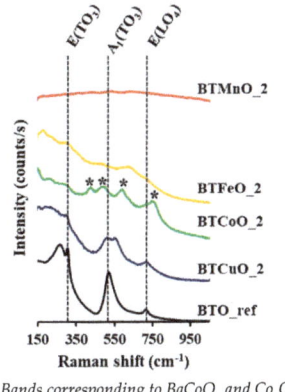

Figure 2. Catalysts characterization: Raman spectra.

2.2. Catalytic Activity

For the analysis of the activity of the catalysts for NO to NO$_2$ oxidation and NOx adsorption/desorption, temperature programmed reaction (TPR-NOx) experiments were carried out. These experiments also allow the selection of the optimal temperature for isothermal NOx storage experiments that have been carried out in order to determine the NOx storage capacity (NSC, which is the amount of NOx stored (in µmol) per gram of catalyst). The results obtained as explained in the Materials and Methods section are presented in Figure 3.

Figure 3a features the NOx conversion profiles for all catalysts. It has been considered that positive values of NOx% conversion indicate that NOx adsorption is taking place, while negative values correspond to a NOx desorption process. According to this, at temperatures lower than 500 °C, approximately, the NOx conversion profiles represent NOx adsorption profiles and, at temperatures higher than 500 °C, these represent NOx desorption profiles. An analysis of the NOx conversion profiles reveals that BTCoO_2 but, mainly, BTCuO_2 catalysts show NOx adsorption/desorption activity, this performance being consistent with the presence of Ba$_2$TiO$_4$ segregated phase, which has been suggested as an active phase for NOx adsorption [11,12].

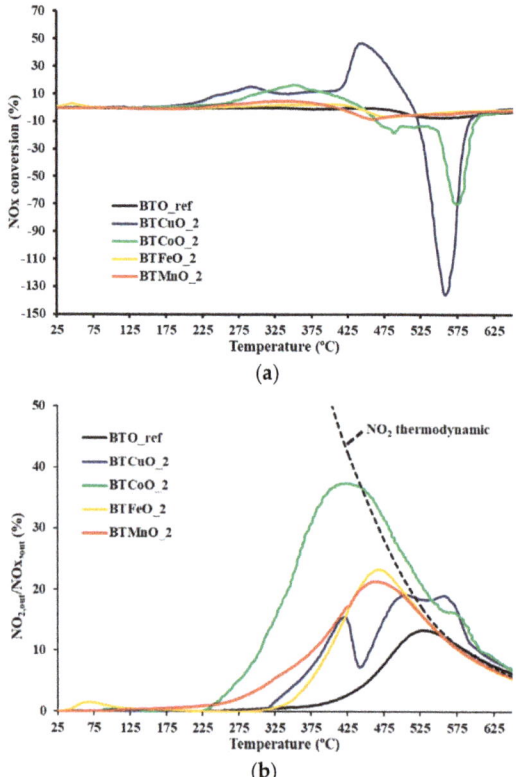

Figure 3. (a) NOx conversion profiles and (b) NO$_2$ generation profiles during the TPR-NOx experiments.

Before analyzing the NO$_2$ generation profiles shown in Figure 3b, it is worth mentioning that NO$_2$ is the main compound involved in NOx adsorption processes [8,11,12] and, for this reason, it has to be considered that the NO$_2$ registered by the analyzers is only the evolved NO$_2$, that is: i) below 500 °C, it is the fraction of NO$_2$ generated which is not stored; and ii) above 500 °C, it represents the NO$_2$ that is being desorbed. Therefore, the NO$_2$ generation shown in Figure 3b cannot be considered as a straight representation of the total NO$_2$ generated and, consequently, any conclusion regarding NO to NO$_2$ oxidation activity of the catalysts must be drawn from the combination of Figure 3a,b.

Thus, all the BaTi$_{0.8}$B$_{0.2}$O$_3$ perovskite catalysts increase the rate of NO$_2$ generation percentage at low temperature as the %NO$_2$ generated is higher than that shown by the BTO perovskite used as a reference (BTO_ref in Figure 3a,b). However, a deeper analysis of the data reveals some significant differences in the NO$_2$ profiles of the catalysts. Firstly, BTFeO_2 and BTMnO_2 catalysts feature Gaussian-shape NO$_2$ generation profiles with maxima at around 470 °C. Considering that almost any significant NOx adsorption/desorption activity is observed for these catalysts, it can be suggested that they are mainly active for NO to NO$_2$ oxidation (however, NOx adsorption capacity cannot be totally ruled out due to the intrinsic characteristics of TPR experiment). Secondly, although BTCoO_2 catalyst presents a similar type of NO$_2$ generation profile, it shows the highest NO oxidation activity, which seems to be related to the presence of Co$_3$O$_4$, as metal oxides are active for the NO to NO$_2$ oxidation reaction [11,12,14–16]. In addition, the low intensity NOx conversion peaks observed for BTCoO_2 catalyst in Figure 3a, indicates a low NOx adsorption activity that, according to literature [14–16], could be due to the presence of minority segregated phases and oxygen vacancies. Finally, a different NO$_2$ generation profile with two maxima and a minimum (at ca. 421, 507, and 441 °C, respectively) are clearly identified for BTCuO_2 catalyst. It is worth indicating that the temperature of the minimum NO_2 generation perfectly matches with the temperature of the maximum NOx conversion observed for this catalyst in Figure 3a. This result points out that, at this temperature, BTCuO_2 shows higher NOx adsorption rate than NO oxidation rate as Ba_2TiO_4 phase, which is active for NOx adsorption [11,12], has been identified.

As a summary, TPR- NOx results reveal that only BTCuO_2 presents the NOx conversion and NO_2 generation profiles expected for LNT catalysts [11,12]. Thus, even though all the catalysts present active sites for NO-to-NO_2 oxidation, such as oxygen vacancies and surface metal oxides, only the catalyst containing copper shows the presence of the Ba_2TiO_4 segregated phase, which is active for NOx adsorption [11,12].

In order to determine the NSC, NOx storage experiments at 400 °C (the minimum temperature for NOx adsorption in TPR-NOx profiles) have been carry out for the three perovskites in which Ti has been substituted in a larger degree, that is, BTCuO_2, BTFeO_2, and BTMnO_2. The NSC values (shown in Table 2) have been obtained during the 10th NOx storage cycle at which the catalysts achieve a stable performance (see Materials and Methods section for more details). Figure 4 shows, as an example, the NO, NO_2, and NOx profiles during NSC experiments at 400 °C, corresponding to the BTCuO_2 catalyst.

Table 2. NSC data at 400 °C for BTO reference, $BaTi_{0.8}B_{0.2}O_3$ catalysts and for some reference noble metal-base catalysts.

Catalyst	NSC (μmol/g)	Temperature (°C)	Lean Cycle Time (s)
BTMnO_2	83	400	300
BTFeO_2	99	400	300
BTCuO_2	269	400	300
1%Pt/20%BaO/Al_2O_3 [10]	150	350	120
2.2%Pt/16.3%BaO/Al_2O_3 [11]	400	350	240
2.2%Pt/20.8%BaO/Al_2O_3 [12]	240	300	240

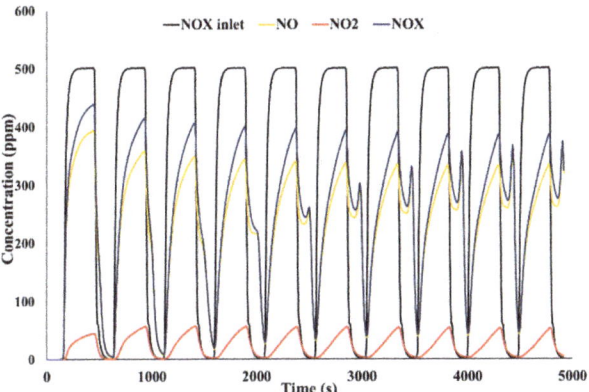

Figure 4. NSC cycles at 400 °C for the BTCuO_2 catalyst.

Data on Table 2 reveals that the three catalysts present a measurable NSC, but, in agreement with TPR-NOx results, BTCuO_2 catalyst is the most active one. The characterization results previously discussed allow us to justify the high NSC shown by copper perovskite. On the one hand, the incorporation of Cu into the perovskite lattice distorts the raw tetragonal structure, generates the largest pool of oxygen vacancies, and promotes the segregation of mainly Ba_2TiO_4 and $BaCO_3$, but also CuO as segregated phases that seem to be the active sites for both NO to NO_2 oxidation and NOx storage [11,12]. On the other hand, the insertion of Mn and Fe causes a structural change from tetragonal to cubic, but only $BaCO_3$ appears as segregated phase and a lower amount of oxygen vacancies respect to Cu. Consequently, a lower NO to NO_2 oxidation activity and NSC is shown by these two catalysts. Finally, it is important to underline that the NSC of BTCuO_2 is within the range of values reported for noble metal/alkali or alkali earth base catalysts (Table 2). Moreover, the BTCuO_2 perovskite does not incorporate any noble metal, and therefore it could be a cheaper alternative to current catalysts based on noble metals. Additionally, as this catalyst works at 400 °C, presenting an acceptable NOx storage capacity, it could be proposed as a component of high-temperature LNT for lean burn gasoline engines (GDI gasoline direct injection) which need catalysts working between 400 and 500 °C.

3. Materials and Methods

Four $BaTi_{0.8}B_{0.2}O_3$ catalysts (being B Mn, Fe, Co, or Cu), named BTMnO_2, BTFeO_2, BTCoO_2, and BTCuO_2 respectively, were prepared by the sol–gel method as previously described [11]. Summarizing, first of all, the hydrolysis of titanium isopropoxide (Ti) was carried out, dissolving the resulting species in an aqueous solution of citric acid (CA) (Ti:CA = 1:2) and hydrogen peroxide (Ti:H_2O_2 = 2:1), and obtaining the citrate–peroxo–titanate (IV) complex. Subsequently, NH_3 was used in order to adjust the pH to 8.5, and the addition of an stoichiometric ($BaTi_{0.8}B_{0.2}O_3$) amount of barium (Ba:Ti = 1:1) and metals precursors (barium acetate and Fe, Co, Cu, and Ni nitrates), took place. During 5 h, until the obtention of a gel, the temperature of the mixture remained 65 °C. Afterwards, a temperature of 90 °C or 24 h was used to dry the sample, which was in the end calcined at 850 °C for 6 h.

To measure the metal content in the samples by ICP-OES, a Perkin-Elmer device model Optima 4300 DV was used. An Autosorb-6B instrument from Quantachrome served to determine, by N_2 adsorption at −196 °C, the BET surface area of the samples. To identify different phases and crystalline structures, X-ray diffraction (XRD) and Raman spectroscopy were employed. XRD tests were performed with a Rigaku Miniflex II powder diffractometer, using Cu Kα (0,15418 nm) radiation with the 2θ angle in the range 20 to 80°, with a step of 0.025° and a time per step of 2 s. Raman scattering spectra were obtained on a Jobin-Ivon dispersive Raman spectrometer (model LabRam) with a variable power He:Ne laser source (633 nm) in the range of 100–1000 nm. To register the XPS spectra, a K-Alpha photoelectron spectrometer by Thermo-Scientific, with an Al Kα (1486.6 eV) radiation source, was used in the following conditions: 5×10^{-10} mbar pressure in the chamber and setting the C1s transition at 284.6 eV, and the binding energy (BE) and kinetic energy (KE) values then determined with the peak-fit software of the spectrophotometer, to regulate the BE and KE scales.

The catalytic activity of the samples (80 mg of catalyst diluted in 300 mg SiC) was tested using two different experiments in a fixed-bed quartz reactor at atmospheric pressure and under a gas flow (500 mL/min): i) temperature programmed reaction (TPR-NOx) tests (10 °C/min, 800 °C) in a gas mixture of 500 ppm NOx and 5 % O_2 and ii) NOx storage cyclic tests at 400 °C, with a gas mixture composed of: i) for lean (storage) cycle (5 min), 500 ppm NOx and 5 % O_2 balanced with N_2, and ii) for rich (regeneration) cycle (3 min), 10% H_2 balanced with N_2. To achieve the stability of the catalysts, and then determine the NSC, 10 consecutive storage–regeneration cycles were accomplished. The gas composition was controlled by specific NDIR-UV gas analyzers for NO, NO_2, CO, CO_2, and O_2 (Rosemount Analytical Model BINOS 1001, 1004, and 100).

NOx conversion profiles as a function of temperature were obtained using the next equation

$$\text{NOx conversion (\%)} = \frac{\text{NOx}_{in} - \text{NOx}_{out}}{\text{NOx}_{in}} \times 100$$

where 'NOx$_{in}$' is the concentration of NOx (=NO + NO_2) feed to the reactor and 'NOx$_{out}$' is the concentration of NOx that leaves the reactor.

The percentage of NO_2 generated during TPR was determined with the equation

$$NO_2(\%) = \frac{NO_{2,out}}{NO_{2,in}} \times 100$$

where 'NO_{2out}' is the concentration of NO_2 that leaves the reactor.

The NSC was obtained as the difference between the NOx signal when the reactor is unfilled and the NOx signal when the reactor is full of catalyst with

$$NOx\ storage = \int_{t0}^{tf} NOx_{inlet}(t) - NOx_{exp}(t)dt$$

where 'NOx_{inlet}' is the concentration of NOx (=NO + NO_2) measured when the reactor is empty, and 'NOx_{exp}' is the concentration of NOx during the NOx storage test.

4. Conclusions

From the analysis of the effect of Ti partial substitution by Mn, Fe, Co, or Cu on the NOx storage capacity (NSC) of the $BaTi_{0.8}B_{0.2}O_3$ lean NOx trap (LNT) catalyst, the following conclusions have been obtained:

- In $BaTi_{0.8}B_{0.2}O_3$ perovskites, Ti is partially substituted by Mn, Fe, Cu and, to a lower extent, by Co.
- The perovskite structure is modified or changed due to the insertion of B metal into the lattice:

    o (i) For the BTCuO_2 catalyst, the tetragonal structure of the raw perovskite is distorted, a larger amount of oxygen vacancies is generated and Ba_2TiO_4 and $BaCO_3$ appear as main minority segregated phases, but CuO is also detected.

    o (ii) For Mn, Fe, and Co, the tetragonal structure changes to cubic, a lower amount of oxygen vacancies are formed and $BaCO_3$ appears as segregated phase in the three catalysts. BTCoO_2 presents also $BaCoO_3$ and Co_3O_4 segregated phases due to a lower degree of Co insertion into the framework.

- Due to the described modifications, all the $BaTi_{0.8}B_{0.2}O_3$ catalysts are active for the NO oxidation to NO_2, which takes place on oxygen vacancies and metal oxide sites, but only the BTCuO_2 catalyst (for which Ba_2TiO_4 segregated phase is identified), presents a significant NOx storage capacity. In fact, at 400 °C, the BTCuO_2 catalyst features the highest NSC which is close to that shown by platinum base catalysts.

Author Contributions: Conceptualization, V.A.-F. and M.-J.I.-G.; Methodology, V.A.-F. and M.-J.I.-G.; Validation, V.A.-F, M.-S.S.-A., and M.-J.I.-G.; Formal analysis, C.A., V.T.-R., V.A.-F, M.-S.S.-A., and M.-J.I.-G.; Investigation, C.A., V.T.-R., V.A.-F.; Resources, M.-S.S.-A. and M.-J.I.-G.; Data curation, V.A.-F., M.-S.S.-A. and M.-J.I.-G.; Writing—original draft preparation, C.A., V.T.-R., and V.A.-F; Writing—review and editing, M.-S.S.-A. and M.-J.I.-G.; Visualization, C.A., V.A.-F., and M.-S.S.-A.; Supervision, V.A.-F. and M.-J.I.-G; Project administration, M.-J.I.-G.; Funding acquisition, M.-S.S.-A. and M.-J.I.-G.

Funding: This research was funded by Generalitat Valenciana (PROMETEO/2018/076 and Ph.D. grant ACIF 2017/221), Spanish Government (MINECO Project CTQ2015-64801-R) and EU (FEDER Founding).

Conflicts of Interest: The authors declare no conflict of interest.

Appendix A

Figure A1 shows the XPS spectra obtained for (a) BTCuO_2, (b) BTCoO_2, (c) BTFeO_2, and (d) BTMnO_2. Each figure contains the XPS spectra for the substituted metal ($Cu2p^{3/2}$, $Co2s$, $Fe2p^{3/2}$, $Mn2p^{3/2}$ transitions), oxygen (O1s transition), and barium ($Ba3d^{5/2}$ transition). In the spectra, the red lines represent the normalized peak and the blue lines represent the deconvolution of the normalized peaks.

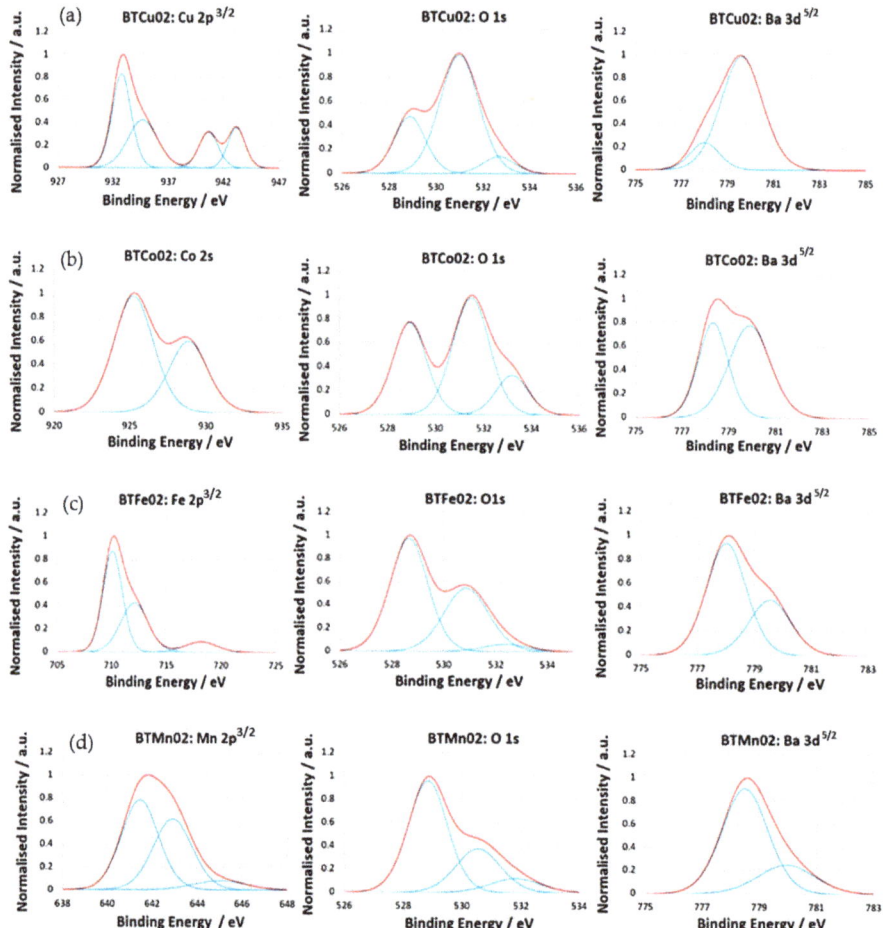

Figure A1. XPS spectra obtained for (**a**) BTCuO_2, (**b**) BTCoO_2, (**c**) BTFeO_2, and (**d**) BTMnO_2.

The XPS spectra of the O1s transition show three contributions for all the catalysts that, according to literature [22,23], can be ascribed to: (i) lattice oxygen of metal oxides at ca. 529 eV; (ii) surface oxygen species such as oxygen peroxides (O_2^{2-}), surface carbonates (CO_3^{2-}), and/or hydroxyl groups (OH^-), at ca. 531 eV; and (iii) adsorbed water at ca. 533 eV.

The XPS spectra of the Ba3d$^{5/2}$ transition for all the catalysts show two contributions at approximately 778 eV and 780 eV binding energies. According to literature [24], they can be ascribed to: (i) Ba(II) oxide species ($BaCuO_2$, $BaTiO_3$-tet., Ba_2TiO_4), and (ii) $BaCO_3$ respectively.

The assignment of the different contributions found in the XPS spectra of the substituted metal (Cu2p$^{3/2}$, Co2s, Fe2p$^{3/2}$, Mn2p$^{3/2}$) transitions is shown in Table A1.

Table A1. Binding energy and assignment of the Cu2p, Co2s, Fe2p, and Mn2p transitions.

XPS Transition	Binding Energy (eV)	Assigned Species
Cu2p$^{3/2}$	932.9	Cu(II) oxide surface species [25,26]
	934.8	Lattice Cu(II) [11]
	940.8	Cu(II) satellite [25]
	943.3	Cu(II) satellite [25]
Co2s	925.28	Co (II) oxide species [24]
	928.88	Co(III) oxide species (Co$_3$O$_4$)
Fe2p$^{3/2}$	710.12	Fe(II) oxide species [24]
	712.09	Fe (III) oxide species [24]
	718.15	Fe (III) Satellite [25]
Mn2p$^{3/2}$	641.46	Mn(III) oxide species (BaMn$_8$O$_{16}$) [27]
	642.90	Mn(IV) oxide species (BaMn$_8$O$_{16}$, BaMnO$_{3-x}$) [27]
	644.98	Mn satellite [26,28]

References

1. Hasan, A.O.; Abu-jrai, A.; Turner, D.; Tsolakis, A.; Xu, H.M.; Golunski, S.E.; Herreros, J.M. Control of harmful hydrocarbon species in the exhaust of modern advanced GDI engines. *Atmos. Environ.* **2016**, *129*, 210–217. [CrossRef]
2. Polat, S.; Uyumaz, A.; Solmaz, H.; Yilmaz, E.; Topgül, T.; Yücesu, H.S. A numerical study on the effects of EGR and spark timing to combustion characteristics and NOx emission of a GDI engine. *Int. J. Green Energy* **2016**, *13*, 63–70. [CrossRef]
3. Epling, W.; Nova, I.; Szanyi, J.; Yezerets, A. Diesel emissions control catalysis. *Catal. Today* **2010**, *151*, 201. [CrossRef]
4. Breen, J.P.; Marella, M.; Pistarino, C.; Ross, J.R.H. Sulfur-Tolerant NOx Storage Traps: An Infrared and Thermodynamic Study of the Reactions of Alkali and Alkaline-Earth Metal Sulfates. *Catal. Lett.* **2002**, *80*, 123–128. [CrossRef]
5. Gill, L.J.; Blakeman, P.G.; Twigg, M.V.; Walker, A.P. The Use of NOx Adsorber Catalysts on Diesel Engines. *Top. Catal.* **2004**, *28*, 157–164. [CrossRef]
6. Pereda-Ayo, B.; González-Velasco, J.R. NOx Storage and Reduction for Diesel Engine Exhaust Aftertreatment. In *Diesel Engine—Combustion Emissions and Condition Monitoring*; Bari, S., Ed.; IntechOpen Ltd.: London, UK, 2013; pp. 161–196. ISBN 978-953-51-1120-7.
7. Bowker, M. Automotive catalysis studied by surface science. *Chem. Soc. Rev.* **2008**, *37*, 2204–2211. [CrossRef] [PubMed]
8. Epling, W.S.; Campell, L.E.; Currier, N.W.; Parks, J.E. Overview of the Fundamental Reactions and Degradation Mechanisms of NOx Storage/Reduction Catalysts. *Catal. Rev. Sci. Eng.* **2004**, *46*, 163–245. [CrossRef]
9. Ad Hoc Working Group on Defining Critical Raw Materials. *Report on Critical Raw Materials for the EU*. 26 May 2014. Available online: http://www.catalysiscluster.eu/wp/wp-content/uploads/2015/05/2014_Critical-raw-materials-for-the-EU-2014.pdf (accessed on 15 January 2019).
10. Royer, S.; Duprez, D.; Can, F.; Courtois, X.; Batiot-Dupeyrat, C.; Laassiri, S.; Aalmdari, H. Perovskites as Substitutes of Noble Metals for Heterogeneous Catalysis: Dream or Reality. *Chem. Rev.* **2014**, *114*, 10292–10368. [CrossRef]
11. Albaladejo-Fuentes, V.; López-Suárez, F.E.; Sánchez-Adsuar, M.S.; Illán-Gómez, M.J. BaTi$_{1-x}$Cu$_x$O$_3$ perovskites: The effect of copper content in the properties and in the NOx storage capacity. *Appl. Catal. A Gen.* **2014**, *488*, 189–199. [CrossRef]
12. Albaladejo-Fuentes, V.; López-Suárez, F.E.; Sánchez-Adsuar, M.S.; Illán-Gómez, M.J. Tailoring the properties of BaTi$_{0.8}$Cu$_{0.2}$O$_3$ catalyst selecting the synthesis method. *Appl. Catal. A Gen.* **2016**, *519*, 7–15. [CrossRef]
13. Modeshia, D.R.; Walton, R.I. Solvothermal synthesis of perovskites and pyrochlores: Crystallisation of functional oxides under mild conditions. *Chem. Soc. Rev.* **2010**, *39*, 4303–4325. [CrossRef]

14. Wang, J.; Su, Y.; Wang, X.; Chen, J.; Zhao, Z.; Shen, M. The effect of partial substitution of Co in LaMnO$_3$ synthesized by sol–gel methods for NO oxidation. *Catal. Commun.* **2012**, *25*, 106–109. [CrossRef]
15. Dhal, G.C.; Dey, S.; Mohan, D.; Prasad, R. Study of Fe, Co, and Mn-based perovskite-type catalysts for the simultaneous control of soot and NOX from diesel engine exhaust. *Mater. Discov.* **2017**, *10*, 37–42. [CrossRef]
16. Li, Z.; Meng, M.; Zha, Y.; Dai, F.; Zhang, J. Highly efficient multifunctional dually-substituted perovskite catalysts La$_{1-x}$K$_x$Co$_{1-y}$Cu$_y$O$_{3-\delta}$ used for soot combustion, NOx storage and simultaneous NOx-soot removal. *Appl. Catal. B Environ.* **2012**, *121–122*, 65–74. [CrossRef]
17. Xia, F.; Liu, J.; Gu, D.; Zhao, P.; Zhang, J.; Che, R. Microwave absorption enhancement and electron microscopy characterization of BaTiO$_3$ nano-torus. *Nanoscale* **2011**, *3*, 3860–3867. [CrossRef]
18. Hayashi, H.; Nakamura, T.; Ebina, T. In-situ Raman spectroscopy of BaTiO$_3$ particles for tetragonal–cubic transformation. *J. Phys. Chem. Solids* **2013**, *74*, 957–962. [CrossRef]
19. Asiaie, R.; Zhu, W.D.; Akbar, S.A.; Dutta, P.K. Characterization of Submicron Particles of Tetragonal BaTiO$_3$. *Chem. Mater.* **1996**, *8*, 226–234. [CrossRef]
20. Torregrosa-Rivero, V.; Albaladejo-Fuentes, V.; Sánchez-Adsuar, M.S.; Illán-Gómez, M.J. Copper doped BaMnO$_3$ perovskite catalysts for NO oxidation and NO$_2$-assisted diesel soot removal. *RSC Adv.* **2017**, *7*, 35228–35238. [CrossRef]
21. Moreno-Marcos, C.; Torregrosa-Rivero, V.; Albaladejo-Fuentes, V.; Sánchez-Adsuar, M.S.; Illán-Gómez, M.J. BaFe$_{1-x}$Cu$_x$O$_3$ Perovskites as Soot Oxidation Catalysts for Gasoline Particulate Filters (GPF): A Preliminary Study. *Top. Catal.* **2018**. [CrossRef]
22. Merino, N.A.; Barbero, B.P.; Eloy, P.; Cadus, L.E. La$_{1-x}$Ca$_x$CoO$_3$ perovskite-type oxides: Identification of surface oxygen species by XPS. *Appl. Surf. Sci* **2016**, *253*, 1489–1493. [CrossRef]
23. Ponce, S.; Peña, M.; Fierro, J.L. Surface properties and catalytic performance in methane combustion of Sr-substituted lanthanum manganites. *Appl. Catal. B Environ.* **2000**, *24*, 193–205. [CrossRef]
24. NIST X-ray Photoelectron Spectroscopy Database. (n.d.). Available online: https://srdata.nist.gov/xps/ (accessed on 15 January 2019).
25. XPS Simplified Knowledge Base. (n.d.). Available online: http://xpssimplified.com/knowledgebase.php (accessed on 15 January 2019).
26. Mingmei, W.; Qiang, S.; Gang, H.; Yufang, R.; Hongyiang, W. Preparation and Properties of BaCuO$_{2.5}$ and Its Related Oxides. *J. Solid State Chem.* **1994**, *110*, 389–392. [CrossRef]
27. Lan, L.; Li, Y.; Zeng, M.; Mao, M.; Ren, L.; Yang, Y.; Liu, H.; Yun, L.; Zhao, X. Efficient UV–vis-infrared light-driven catalytic abatement of benzene on amorphous manganese oxide supported on anatase TiO$_2$ nanosheet with dominant {001} facets promoted by a photothermocatalytic synergetic effect. *Appl. Catal. B Environ.* **2017**, *203*, 494–504. [CrossRef]
28. Liu, P.; He, H.; Wei, G.; Liu, D.; Liang, X.; Chen, T.; Zhu, J.; Zhu, R. An efficient catalyst of manganese supported on diatomite for toluene oxidation: Manganese species, catalytic performance, and structure-activity relationship. *Microporous Mesoporous Mater.* **2017**, *239*, 101–110. [CrossRef]

© 2019 by the authors. Licensee MDPI, Basel, Switzerland. This article is an open access article distributed under the terms and conditions of the Creative Commons Attribution (CC BY) license (http://creativecommons.org/licenses/by/4.0/).

Article

Catalytic Oxidation of NO over LaCo$_{1-x}$B$_x$O$_3$ (B = Mn, Ni) Perovskites for Nitric Acid Production

Ata ul Rauf Salman [1], Signe Marit Hyrve [1], Samuel Konrad Regli [1], Muhammad Zubair [1], Bjørn Christian Enger [2], Rune Lødeng [2], David Waller [3] and Magnus Rønning [1,*]

[1] Department of Chemical Engineering, Norwegian University of Science and Technology (NTNU), Sem Sælands vei 4, NO-7491 Trondheim, Norway; ata.r.salman@ntnu.no (A.u.R.S.); signemhyrve@gmail.com (S.M.H.); samuel.k.regli@ntnu.no (S.K.R.); muhammad.zubair@ntnu.no (M.Z.)
[2] SINTEF Industry, Kinetic and Catalysis Group, P.O. Box 4760 Torgarden, NO-7465 Trondheim, Norway; bjorn.christian.enger@sintef.no (B.C.E.); rune.lodeng@sintef.no (R.L.)
[3] YARA Technology Center, Herøya Forskningspark, Bygg 92, Hydrovegen 67, NO-3936 Porsgrunn, Norway; david.waller@yara.com
* Correspondence: magnus.ronning@ntnu.no; Tel.: +47-73594121

Received: 31 March 2019; Accepted: 3 May 2019; Published: 8 May 2019

Abstract: Nitric acid (HNO$_3$) is an important building block in the chemical industry. Industrial production takes place via the Ostwald process, where oxidation of NO to NO$_2$ is one of the three chemical steps. The reaction is carried out as a homogeneous gas phase reaction. Introducing a catalyst for this reaction can lead to significant process intensification. A series of LaCo$_{1-x}$Mn$_x$O$_3$ (x = 0, 0.25, 0.5 and 1) and LaCo$_{1-y}$Ni$_y$O$_3$ (y = 0, 0.25, 0.50, 0.75 and 1) were synthesized by a sol-gel method and characterized using N$_2$ adsorption, ex situ XRD, in situ XRD, SEM and TPR. All samples had low surface areas; between 8 and 12 m^2/g. The formation of perovskites was confirmed by XRD. The crystallite size decreased linearly with the degree of substitution of Mn/Ni for partially doped samples. NO oxidation activity was tested using a feed (10% NO and 6% O$_2$) that partly simulated nitric acid plant conditions. Amongst the undoped perovskites, LaCoO$_3$ had the highest activity; with a conversion level of 24.9% at 350 °C; followed by LaNiO$_3$ and LaMnO$_3$. Substitution of LaCoO$_3$ with 25% mol % Ni or Mn was found to be the optimum degree of substitution leading to an enhanced NO oxidation activity. The results showed that perovskites are promising catalysts for NO oxidation at industrial conditions.

Keywords: NO oxidation; catalytic oxidation; nitric oxide; perovskite; nitric acid; ostwald's process; in situ; LaCoO$_3$; LaMnO$_3$; LaNiO$_3$

---

1. Introduction

Oxidation of nitric oxide (Equation (1)) is one of the few known third order reactions. The reaction is unusual, as the rate of reaction increases with a decrease in the temperature [1].

$$2NO + O_2 \leftrightharpoons NO_2 \qquad \Delta_r H_{298} = -113.8 \text{ kJ/mol} \qquad (1)$$

NO oxidation is a key reaction in lean NO$_x$ abatement technologies and in the Ostwald process for nitric acid production. In Ostwald's process, NO oxidation is carried out as a non-catalytic process and the forward reaction is favored by the removal of heat and by providing sufficient residence time. Typical gas stream concentrations are 10% NO, 6% O$_2$ and 15% H$_2$O [2]. Using a catalyst for NO oxidation may lead to significant process intensification of the nitric acid plant. In addition to speeding up the oxidation process, it may reduce capital costs and increase heat recovery. Efforts have been made to find a catalyst effective under industrial conditions; but success so far has not been achieved [2]. To date, the process is carried out as a homogenous process in modern nitric acid plants. To the best of the authors knowledge, apart from an earlier patent [3], only two recent studies [4,5] report catalytic oxidation of NO at nitric acid plant conditions. However, catalytic oxidation of NO has been extensively studied with regards to lean NO_x abatement technologies and reviewed by Russel and Epling [6] and Hong et al. [7]. In these studies, oxidation of NO is carried out at very lean concentrations of NO ranging from 100–1000 ppm of NO [8–10]. The huge difference in NO concentration between NO_x abatement and nitric acid production makes the extrapolation of these findings to nitric acid plant conditions, questionable.

Although it has been demonstrated that platinum has significant catalytic activity for oxidation of NO to NO_2 at nitric acid plant conditions [4,5] the high-cost and scarcity of platinum motivates the search for potential non-noble metal based catalysts.

Perovskites have gained particular interest as catalytic materials due to their high thermal stability, ease of synthesis and good catalytic activity [11]. Perovskites are represented by a general formula ABO_3, where A represents a rare earth or alkaline earth cation and B represents a transition metal cation. The activity of perovskites can be tuned by partial substitution of A and/or B site cations with another element to obtain the desired properties [11]. The lanthanum-based perovskite *$LaBO_3$* (B = Co, Mn, Ni) have been found to be active for NO oxidation at lean NO conditions [12,13]. It has been demonstrated for $LaCoO_3$, that partial doping of the A-site with strontium or cerium or doping of the B-site with manganese or nickel enhances NO oxidation activity of the perovskite [13–16].

In this work, lanthanum-based perovskites with cobalt, nickel and manganese ($LaCoO_3$, $LaNiO_3$ and $LaMnO_3$) were synthesized by the sol-gel method using citric acid. Catalytic tests were performed using a dry feed (10% NO, 6% O_2) at atmospheric pressure; partially simulating nitric acid plant conditions. The effect of B-site substitution was studied by preparing a series of $LaCo_{1-x}Mn_xO_3$ and $LaCo_{1-y}Ni_yO_3$ catalysts. The catalysts were characterized using N_2 adsorption, X-ray diffraction (XRD), Scanning electron microscopy (SEM) and temperature programmed reduction (TPR). Catalyst structure during pretreatment and oxidation of NO was monitored using in situ XRD.

2. Results and Discussion

2.1. Catalyst Characterisation

Specific surface areas are summarized in Table 1. All samples have a relatively low surface area, in the range of 8–12 m^2/g, which is a typical of perovskites prepared by the citrate method [14,17,18]. The surface area remains unaffected by the degree of substitution of Mn (x). However, an irregular change in surface area is observed with the degree of substitution of Ni (y). For $y = 0.25$ and 0.75, the surface area increased by 25% (from 8 to 10 m^2/g) and 50% (from 8 to 12 m^2/g) respectively. However, for $y = 0.50$ the surface area remains the same as for $LaCoO_3$.

Table 1. Brunauer–Emmett–Teller (BET) surface area and crystallite size (d) calculated using the Scherrer equation.

Sample	Surface Area (m^2/g)	d (nm)
$LaCoO_3$	8	28
$LaCo_{0.75}Mn_{0.25}O_3$	8	24
$LaCo_{0.50}Mn_{0.50}O_3$	7	18
$LaMnO_3$	9	21
$LaCo_{0.75}Ni_{0.25}O_3$	10	22
$LaCo_{0.50}Ni_{0.50}O_3$	8	17
$LaCo_{0.25}Ni_{0.75}O_3$	12	8
$LaNiO_3$	8	13

The XRD patterns shown in Figure 1 reveal the formation of phase-pure perovskite structure for all samples, and no peaks characteristic of the metallic oxides or carbonates were seen. The main characteristic peak at 2θ = 33° for LaCoO$_3$ slightly shifts towards lower 2θ values with an increase in x and y, indicating an increase in the lattice parameters confirming previous findings [13,19,20]. This expansion can be explained by comparing the ionic radii of the species in the perovskite structure. The higher ionic radii of Ni^{3+} (0.56 Å) than Co^{3+} (0.52 Å) responsible for lattice expansion in LaCo$_{1-y}$Ni$_y$O$_3$ [20]. In case of LaCo$_{1-x}$Mn$_x$O$_3$, the average trivalent metal site is conserved by adjusting the ratio between Mn^{4+}/Mn^{3+} and Co^{3+}/Co^{2+} [21]. Therefore, the relative amount of Mn^{4+} (0.52 Å) present, Mn^{3+} (0.645 Å), Co^{2+} (0.82 Å), Co^{3+} (0.52 Å) dictates the overall lattice parameters.

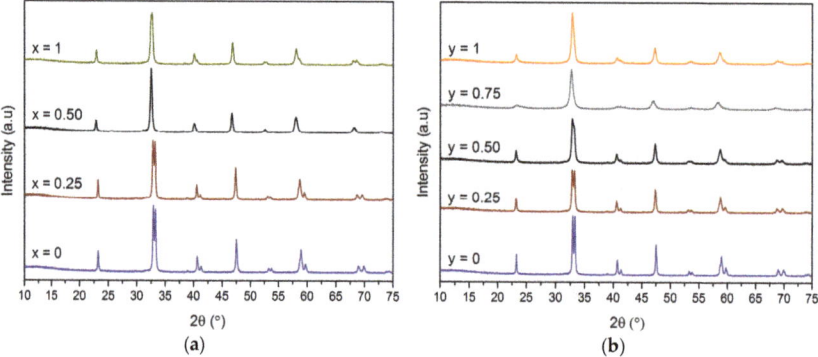

Figure 1. XRD patterns of: (a) LaCo$_{1-x}$Mn$_x$O$_3$; (b) LaCo$_{1-y}$Ni$_y$O$_3$ perovskites.

Among undoped perovskites, LaCoO$_3$ and LaMnO$_3$ belong to the rhombohedral phase in agreement with the results reported in literature [22,23]. LaNiO$_3$ belongs to the cubic phase, which is consistent with previous studies [24].

Crystallite sizes calculated using the Scherrer equation are summarized in Table 1. Figure 2 shows crystallite size as a function of the degree of substitution. The highest crystallite size of 28 nm was observed for LaCoO$_3$ and a linear decrease was observed with an increase in x and y for partially substituted samples indicating that partial substitution effectively restrains the crystal growth. However, other contributions to XRD peak broadening such as strain cannot be ruled out. Partially substituted nickel perovskites had smaller crystallite size compared to their manganese counterparts.

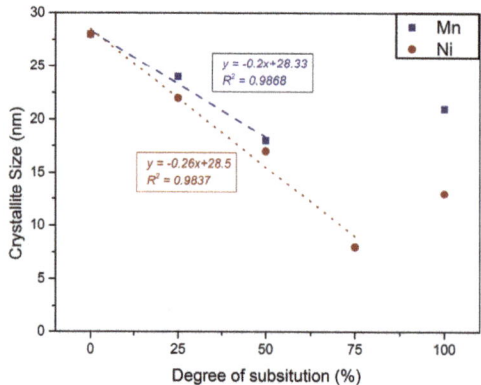

Figure 2. Crystallite size as a function of the degree of substitution for LaCo$_{1-x}$Mn$_x$O$_3$ and LaCo$_{1-y}$Ni$_y$O$_3$ perovskites.

Figure 3 shows the XRD pattern recorded in situ during pretreatment of LaCoO$_3$ and LaMnO$_3$. No change in structure is observed apart from lattice expansion with an increase in temperature. The structural stability of the perovskites is in accordance with the fact that they are calcined at higher temperatures in comparison with the pretreatment temperature of 500 °C. No change in structure was observed for LaCoO$_3$ and LaMnO$_3$ during steady state oxidation of NO at 350 °C indicating that the bulk structure of perovskite remains unaffected during the catalytic process. However, minor changes beyond the detectable range of XRD cannot be ruled out.

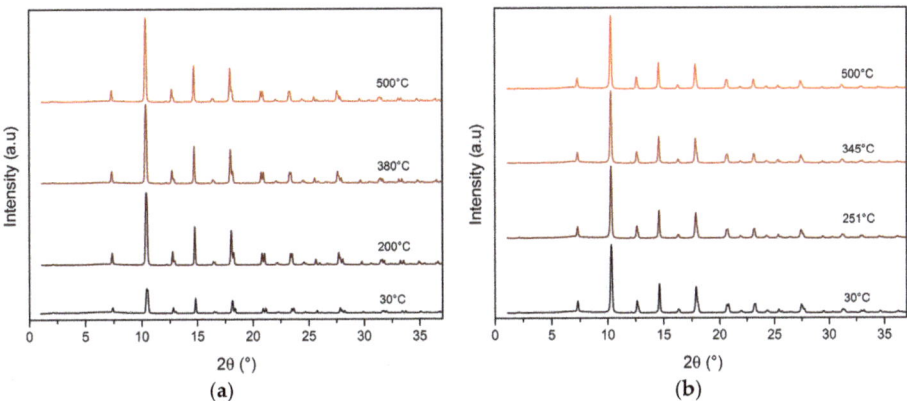

Figure 3. XRD patterns (λ = 0.49324 Å) recorded in situ during pretreatment of: (**a**) LaCoO$_3$; (**b**) LaMnO$_3$ perovskites.

Figure 4 shows the SEM images of LaCoO$_3$ with varying content of Mn and pure LaNiO$_3$. The presence of agglomerated non-spherical particles is observed for all samples. A significant change in morphology is observed with the substitution of Co with Mn along with an increase in the extent of agglomeration (Figure 4b–d).

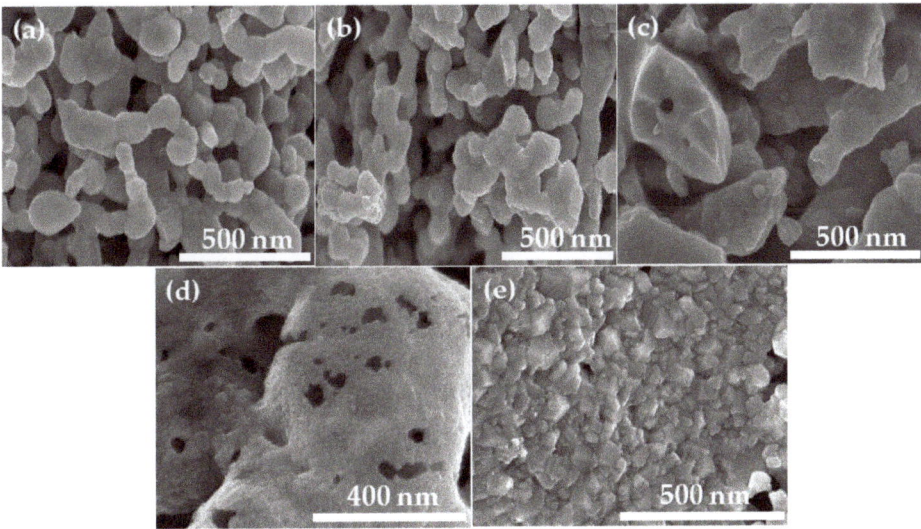

Figure 4. SEM images: (**a**) LaCoO$_3$; (**b**) LaCo$_{0.75}$Mn$_{0.25}$O$_3$; (**c**) LaCo$_{0.50}$Mn$_{0.50}$O$_3$; (**d**) LaMnO$_3$; (**e**) LaNiO$_3$.

The H_2-TPR profiles of $LaCo_{1-x}Mn_xO_3$ perovskites are given in Figure 5a. Three reduction peaks at 334, 373 and 526 °C are observed in the TPR profile of $LaCoO_3$. The first two peaks are attributed to the reduction of Co^{3+} to Co^{2+}, while the peak at 526 °C represents the reduction of Co^{2+} to Co^0 leading to the destruction of the perovskite structure [25]. TPR of $LaMnO_3$ shows two main reduction peaks at 383 and 818 °C. The first peak represents the reduction of Mn^{4+} to Mn^{3+}, while the reduction of Mn^{3+} to Mn^{2+} occurs at elevated temperatures (above 700 °C), forming MnO and simultaneous collapse of the perovskite structure [26]. For $x = 0.25$ and 0.5, broad peaks overlapping reduction peaks of Co^{3+} to Co^{2+} and Mn^{4+} to Mn^{3+} are observed below 550 °C.

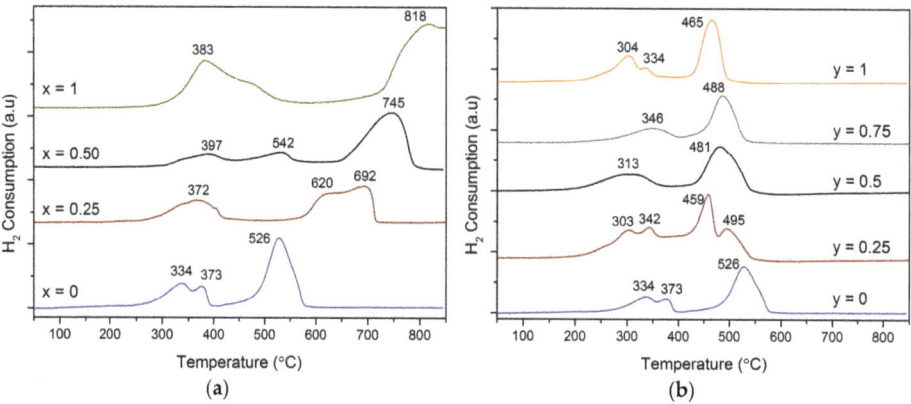

Figure 5. TPR profiles of: (**a**) $LaCo_{1-x}Mn_xO_3$; (**b**) $LaCo_{1-y}Ni_yO_3$ perovskites.

Figure 5b shows H_2-TPR profiles of $LaCo_{1-y}Ni_yO_3$ perovskites. Reduction of $LaNiO_3$ also proceeds via three peaks at 304 °C, one at 334 °C and one at 465 °C. A similar three-step reduction process for $LaNiO_3$ has been reported [27] and in situ XRD revealed that reduction proceeds via formation of the La_2NiO_4 phase [27]. The highest peak is associated with the formation of Ni^0 and La_2O_3 resulting in the destruction of the perovskite structure. Four reduction peaks were observed for $y = 0.25$. Comparison with $LaCoO_3$ and $LaNiO_3$ indicates that the first two peaks at 303 and 342 °C match with the reduction peaks of Ni^{3+} to Ni^{2+} and Co^{3+} to Co^{2+}. Whereas the former two peaks at 459, 495 °C corresponds to further reduction to form the metallic phases (Ni^0 and Co^0). This reduction profile matches well with the previous findings [20]. In contrast to distinct reduction peaks for cobalt and nickel for $LaCo_{0.75}Ni_{0.25}O_3$, only two broad peaks were observed for $y = 0.50$ and 0.75. The first peak overlaps with the reduction of Ni^{3+} to Ni^{2+} and Co^{3+} to Co^{2+} and increases from 313 °C for $y = 0.50$ to 346 °C for $y = 0.75$. Simultaneous reduction to metallic phases (Ni^0 and Co^0) was observed at 418 °C and 488 °C for $y = 0.50$ and 0.75, respectively.

2.2. NO Oxidation Activity

The oxidation of NO occurs as a homogeneous gas-phase reaction with a second order dependency in NO concentration. For this reason, contributions from gas phase conversion have been detected for studies performed at lean NO concentrations [28–30]. This necessitates the quantification of the gas phase contribution to NO oxidation in the current study, which involves such a high concentration of NO (10%). The blank run performed without catalyst is given in our previous publication [5]. The gas phase conversion gradually decreases with increasing temperature in the studied temperature range (150–450 °C); with a conversion of 6.1% at 350 °C.

The conversion of NO over $LaCo_{1-x}Mn_xO_3$ as a function of temperature is shown in Figure 6. The NO conversion and rate of reaction are summarized in Table 2. At low temperature, only gas phase conversion is observed for all perovskites. The catalytic activity for $LaCoO_3$ starts at about 270 °C and increases gradually until it becomes limited by the thermodynamic equilibrium. This is in contrast to studies performed at lean NO_x conditions, where the catalytic activity starts at significantly lower temperatures (150–200 °C); increases with a steeper slope, and becomes thermodynamically limited at ca 300 °C [14,15]. The catalytic conversion starts at a lower temperature (240 °C) for $x = 0.25$ in comparison to other perovskites and a significant increase in conversion was observed. A further increase in x to 0.5 leads to a substantial decrease in NO conversion to conversion levels even lower than what is observed for $LaMnO_3$. Manganese is stable in valence states +III and +IV while cobalt is stable in valence states +II and +III. Ghiasi et al. [21] used X-ray absorption spectroscopy (XAS) to study the valence state of Mn and Co in a series of $LaCo_{1-x}Mn_xO_3$ perovskites and found that the average trivalent metal site is conserved by shifting the balance between Mn^{4+}/Mn^{3+} in combination with Co^{3+}/Co^{2+}. The ratio between Mn^{4+}/Mn^{3+} and Co^{3+}/Co^{2+} decreases with a sequential increase in manganese content, the highest value being observed for $LaCo_{0.75}Mn_{0.25}O_3$. The best activity of $LaCo_{0.75}Mn_{0.25}O_3$ in the series of $LaCo_{1-x}Mn_xO_3$ perovskites may be attributed to the presence of highest content of Mn^{4+} as amongst different valence states of manganese, Mn^{4+} (MnO_2) exhibits the highest activity for NO oxidation followed by Mn^{3+} (Mn_2O_3) and Mn^{2+} (Mn_3O_4) [31].

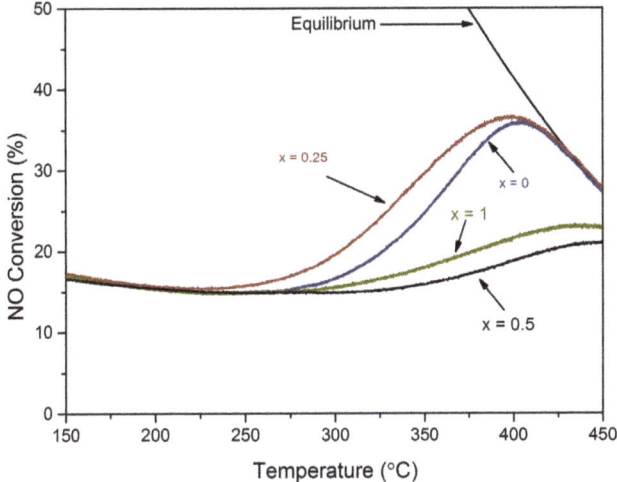

Figure 6. Conversion of NO over $LaCo_{1-x}Mn_xO_3$ as a function of temperature.

Table 2. NO oxidation activity of perovskites at 350 °C.

Sample	X_{NO} (%)	r_{NO} [1] ($\mu mol g_{cat}^{-1} s^{-1}$)
$LaCoO_3$	24.9	5.14
$LaCo_{0.75}Mn_{0.25}O_3$	29.7	6.55
$LaCo_{0.50}Mn_{0.50}O_3$	15.9	2.72
$LaMnO_3$	18.0	3.27
$LaCo_{0.75}Ni_{0.25}O_3$	30.3	6.70
$LaCo_{0.50}Ni_{0.50}O_3$	23.2	4.74
$LaCo_{0.25}Ni_{0.75}O_3$	16.4	2.86
$LaNiO_3$	20.8	4.05

[1] Catalytic activity obtained by subtracting the gas phase conversion of NO.

Figure 7 shows the conversion of NO over $LaCo_{1-y}Ni_yO_3$ as a function of temperature. Minor difference in gas phase conversion is observed at lower temperatures due to the differences in the packing of the catalyst bed. The substitution with 25 mol% Ni in $LaCoO_3$ had a significant positive impact on the activity exhibiting a conversion of 30% at 350 °C. With a further increase in y, the activity gradually decreased until 75% nickel content. Similar conversion curves are exhibited by $y = 0.5$ and $y = 1$ perovskites. Although our results differ from Zhong et al. [13], who reported 70 mol% nickel substitution in LaCoO$_3$ to yield the best results for NO oxidation, it can be argued that they used a co-precipitation method for preparation of perovskites and the activity was tested at substantially different feed concentration (400 ppm NO and 6% O$_2$) compared to the present study. The increase in activity for LaCo$_{0.75}$Ni$_{0.25}$O$_3$ can in part be attributed to the 25% increase in surface area. However, for LaCo$_{0.25}$Ni$_{0.75}$O$_3$ the surface area increased with 50% compared to LaCoO$_3$ and 33% compared to LaNiO$_3$, while the catalytic activity decreased. Thus, it seems likely that the change in catalytic activity is more related to the chemical oxygen dynamics and to the redox properties than to changes in the surface area in this range. Ivanova et al. [32] reported a maximum in defect structures for $x = 0.25$ in a series of LaCo$_{1-x}$Ni$_x$O$_3$, detected by EPR due to the presence of magnetic Ni clusters. However, this depends on the method of preparation and pre-treatment procedure. The highest activity of LaCo$_{0.75}$Ni$_{0.25}$O$_3$ may thus be related to lattice defects not detectable by bulk XRD. The NO oxidation activity follows the order LaCoO$_3$ > LaNiO$_3$ > LaMnO$_3$ for the undoped perovskites.

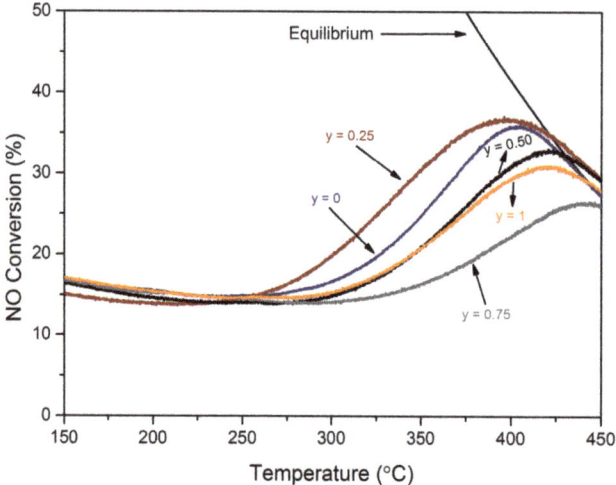

Figure 7. Conversion of NO over LaCo$_{1-y}$Ni$_y$O$_3$ as a function of temperate.

Figure 8 shows the rate of reaction as a function of crystallite size for LaCo$_{1-x}$Mn$_x$O$_3$ and LaCo$_{1-y}$Ni$_x$O$_3$ perovskites. The rate of reaction increases linearly with crystallite size with $x = 0.25$ and $y = 0.25$ samples being the exception and not included in the linear fit. It should be kept in mind that the crystallite sizes were calculated using the Scherrer equation assuming spherical geometry. Though, the crystallites are not spherical as revealed through SEM images. The crystallite size estimates from the Scherrer equation might not be 100% accurate but they provide a fair comparison.

Partial substitution of LaCoO$_3$ with either manganese or nickel leads to a modification in the redox properties, morphology, structure, crystallite size and valence state of the metal site. Therefore, it is difficult to pinpoint one governing factor determining the catalytic activity. In fact, it is a combination of several factors which contribute to dictating catalytic activity.

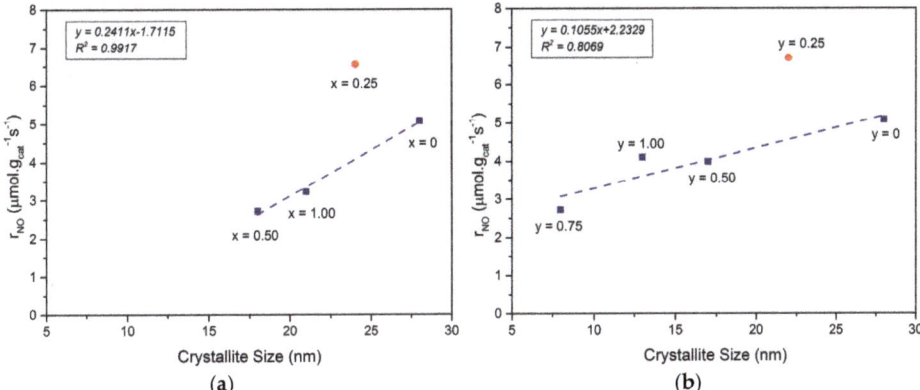

Figure 8. Rate of reaction as a function of crystallite size: (a) $LaCo_{1-x}Mn_xO_3$; (b) $LaCo_{1-y}Ni_yO_3$ perovskites.

3. Materials and Methods

3.1. Catalyst Preparation

All catalysts were prepared using the citrate method [14]. Nitrate salts were used as starting materials: $La(NO_3)_3 \cdot 6H_2O$ (Alfa Aesar, 99.9%, Kandel, Germany), $Mn(NO_3)_2 \cdot 4H_2O$ (Alfa Aesar, 99.98%, Kandel, Germany), $Co(NO_3)_3 \cdot 6H_2O$ (Acros Organics, 99%, Geel, Belgium) and $Ni(NO_3)_2 \cdot 6H_2O$ (Acros Organics, 99%, Geel, Belgium). An appropriate amount of nitrate salts of the desired A and B sites were dissolved in deionized water with 10 wt.% excess citric acid (Sigma Aldrich, 99.5%, Munich, Germany). The solution was stirred for 1 h at room temperature and further stirred in an oil bath at 80 °C until a viscous gel was obtained. The gel was dried in static air overnight at 90 °C and then heated to 150 °C for 1 h. The black, spongy material was crushed and calcined at 700 °C for 5 h in static air.

3.2. Catalyst Characterisation

N_2 adsorption was used to measure the specific surface area. The sample (200 mg) was degassed at 200 °C overnight in VacPrep 061 Degasser (Norcross, GA, USA). Nitrogen adsorption was performed with a Miromeritics TriStar II 3020 Surface Area and Porosity Analyzer (Norcross, GA, USA) at −196 °C. The specific surface areas were calculated using the BET desorption branch of the isotherm [33].

Ex situ XRD patterns of as prepared samples were obtained with a Bruker D8 Advanced X-ray Diffractometer (Cambridge, United Kingdom) with a copper anode with an X-ray wavelength of 1.5418 Å. The measured angles (2θ) were scanned from 10° to 75° in 30 min; at a fixed divergence angle of 0.2°. The PDF database was used for phase identification. The average crystallite sizes were calculated according to the Scherrer equation [34] assuming spherical crystals (K = 0.9) using the diffraction peak between 2θ = 45–50°.

In situ powder XRD experiments were carried out at BM31 of the Swiss-Norwegian beam lines (SNBL) at the European Synchrotron Radiation Facility (ESRF). Catalyst sample (30mg, sieve fraction 53–90 μm) was fixed between two quartz wool plugs in a quartz capillary of 1 mm internal diameter (bed length: 10 mm). The capillary was then mounted in a custom cell [35] and exposed to X-rays for diffraction measurements. The temperature of the capillary reactor was controlled by a calibrated hot air blower. Powder X-ray diffraction patterns were collected with a 2D plate detector (Mar-345) using monochromatic radiation of wavelength 0.49324Å. Note that is different from the Cu Kα wavelength, which is used for acquiring ex situ XRD patterns. The instrumental peak broadening, wavelength calibration, and detector distance corrections were performed using a NIST 660a LaB_6 standard. Mass flow controllers were used to feed NO, O_2 and He in to produce the desired gas-feed composition (15 Ncm^3/min; 1% NO, 6% O_2, He balance). XRD patterns were recorded in situ during the pretreatment step (flowing 15 Ncm^3/min of 6% O_2/He by heating at 10 °C/min from ambient to 500 °C and holding for 1 hr) and steady state NO oxidation (1%NO, 6% O_2 and balance He) at 350 °C for 2 h. A lower concentration of NO (1%) was used for the in situ studies due to experimental constraints at the beamline.

The morphological analysis of as-synthesized perovskites was performed by using an in-lens cold field emission electron microscope FE-S(T)EM, (Hitachi S-5500) in scanning electron microscopy (SEM) mode.

Temperature programmed reduction by H_2 was performed with an Altamira BenchCAT Hybrid 1000 HP (Pittsburgh, PA, USA). The samples (100 mg) were pretreated in 50 Ncm^3/min flow of Ar at 150 °C for 30 min, with a heating rate of 10 °C. TPR was conducted by heating at a rate of 5 °C/min from 50 to 850 °C with a 50 Ncm^3/min flow of 10% H_2/Ar.

3.3. Catalyst Activity Testing

Details of the reactor and experimental setup can be found in our previous publication [5]. Briefly, the activity of the catalysts was measured in a vertical stainless steel tubular reactor (internal diameter = 9.7 mm) operated at atmospheric pressure. The temperature was measured and controlled by a thermocouple, protected by a stainless steel jacket, inserted into the catalyst bed. A MKS MultiGas 2030-HS FTIR Gas Analyzer (path length 5.11 m, Cheshire, United Kingdom), calibrated at 1 bar and 191°C was used to analyze the composition of NO and NO_2 in the product stream.

For activity measurements, 0.5 g of catalyst diluted with 2.75 g SiC was loaded into the reactor and held in place by quartz wool plugs. Prior to the activity tests, the catalysts were pretreated at 500 °C for 1 h in 200 Ncm^3/min flow of 10% O_2/Ar and subsequently cooled down in inert argon atmosphere. The activity of the catalysts was investigated by heating from 150 to 450 °C at a rate of 5 °C/min under a flow of 200 Ncm^3/min of feed gas (10% NO, 6% O_2 in balance Ar).

Conversion of NO to NO_2 was calculated by the following equation:

$$\text{NO conversion} = \alpha \times [NO_2]_{outlet}/[NO]_{inlet} \qquad (2)$$

where $[NO]_{inlet}$ and $[NO_2]_{outlet}$ are concentrations of NO at inlet and NO_2 at the outlet of the reactor. Volume changes arising from the reaction is taken into account by the constant "α" [36] where $\alpha = 0.99$. The closure of nitrogen balance across the reactor (99.5–100%) confirmed that all nitrogen is present as NO and NO_2. Comparison of catalyst activity is performed at 350 °C where the reaction is in the kinetic regime, away from equilibrium. The reaction rate (r_{NO}) calculations were performed by subtracting the homogeneous gas phase conversion of NO; hence, reflecting only the activity provided by the catalyst.

4. Conclusions

A series of lanthanum-based perovskites have been investigated for oxidation of NO using a feed containing 10% NO and 6% O_2, thus partially simulating nitric acid plant conditions. Among the undoped perovskites, the NO oxidation activity follows the order $LaCoO_3$ > $LaNiO_3$ > $LaMnO_3$. A significant increase in NO oxidation activity was achieved by partial substitution of cobalt in $LaCoO_3$ with 25 mol% of either nickel or manganese. Further increase in the degree of Co substitution had a negative impact on activity.

From this work, perovskites are shown to be promising catalysts for oxidizing NO to NO_2 at conditions representative of nitric acid plant operation. Low cost, ease of production and significant catalytic activity make perovskites attractive candidates as alternatives to noble metal catalysts.

Author Contributions: Conceptualization, A.u.R.S., D.W., B.C.E., R.L. and M.R.; methodology, A.S., S.M.H.; formal analysis, A.u.R.S., S.M.H., S.K.R. and M.Z.; investigation, A.u.R.S., S.M.H., S.K.R. and M.Z.; writing—original draft preparation, A.u.R.S.; writing—review and editing, S.M.H., S.K.R, M.Z., B.C.E., R.L, D.W. and M.R.; visualization, A.u.R.S.; supervision, B.C.E., R.L., D.W. and M.R.; project administration, M.R.; funding acquisition, B.C.E., R.L., D.W. and M.R.

Funding: This research was funded by iCSI (industrial Catalysis Science and Innovation) Centre for Research-based Innovation, which receives financial support from the Research Council of Norway, grant number 237922. The Research Council of Norway is also acknowledged for financial support to the Swiss-Norwegian Beamlines at ESRF, grant number 273608.

Conflicts of Interest: The authors declare no conflict of interest.

References

1. Baulch, D.; Drysdale, D.; Horne, D. *Homogeneous Gas Phase Reactions of the H_2-N_2-O_2 System*; CRC Press: Boca Raton, FL, USA, 1973; pp. 285–300.
2. Honti, G. *The Nitrogen Industry*; Akademiai Kiado: Budapest, Hungary, 1976; pp. 400–413.
3. Klingelhoefer, W.C. Nitric Oxide Oxidation. US2115173 A, 26 April 1938.
4. Grande, C.A.; Andreassen, K.A.; Cavka, J.H.; Waller, D.; Lorentsen, O.-A.; Øien, H.; Zander, H.-J.; Poulston, S.; García, S.; Modeshia, D. Process intensification in nitric acid plants by catalytic oxidation of nitric oxide. *Ind. Eng. Chem. Res.* **2018**, *57*, 10180–10186. [CrossRef]
5. Salman, A.u.R.; Enger, B.C.; Auvray, X.; Lødeng, R.; Menon, M.; Waller, D.; Rønning, M. Catalytic oxidation of NO to NO_2 for nitric acid production over a Pt/Al_2O_3 catalyst. *Appl. Catal. A Gen.* **2018**, *564*, 142–146. [CrossRef]
6. Russell, A.; Epling, W.S. Diesel oxidation catalysts. *Catal. Rev.* **2011**, *53*, 337–423. [CrossRef]
7. Hong, Z.; Wang, Z.; Li, X. Catalytic oxidation of nitric oxide (NO) over different catalysts: An overview. *Catal. Sci. Technol.* **2017**, *7*, 3440–3452. [CrossRef]
8. Auvray, X.; Olsson, L. Stability and activity of Pd-, Pt- and Pd–Pt catalysts supported on alumina for NO oxidation. *Appl. Catal. B Environ.* **2015**, *168–169*, 342–352. [CrossRef]
9. Auvray, X.; Pingel, T.; Olsson, E.; Olsson, L. The effect gas composition during thermal aging on the dispersion and NO oxidation activity over Pt/Al_2O_3 catalysts. *Appl. Catal. B Environ.* **2013**, *129*, 517–527. [CrossRef]
10. Després, J.; Elsener, M.; Koebel, M.; Kröcher, O.; Schnyder, B.; Wokaun, A. Catalytic oxidation of nitrogen monoxide over Pt/SiO_2. *Appl. Catal. B Environ.* **2004**, *50*, 73–82. [CrossRef]
11. Zhu, J.; Thomas, A. Perovskite-type mixed oxides as catalytic material for NO removal. *Appl. Catal. B Environ.* **2009**, *92*, 225–233. [CrossRef]
12. Onrubia, J.A.; Pereda-Ayo, B.; De-La-Torre, U.; González-Velasco, J.R. Key factors in Sr-doped $LaBO_3$ (B=Co or Mn) perovskites for NO oxidation in efficient diesel exhaust purification. *Appl. Catal. B Environ.* **2017**, *213*, 198–210. [CrossRef]
13. Zhong, S.; Sun, Y.; Xin, H.; Yang, C.; Chen, L.; Li, X. NO oxidation over Ni–Co perovskite catalysts. *Chem. Eng. J.* **2015**, *275*, 351–356. [CrossRef]
14. Kim, C.H.; Qi, G.; Dahlberg, K.; Li, W. Strontium-doped perovskites rival platinum catalysts for treating NO_x in simulated diesel exhaust. *Science* **2010**, *327*, 1624–1627. [CrossRef]
15. Wen, Y.; Zhang, C.; He, H.; Yu, Y.; Teraoka, Y. Catalytic oxidation of nitrogen monoxide over $La_{1-x}Ce_xCoO_3$ perovskites. *Catal. Today* **2007**, *126*, 400–405. [CrossRef]
16. Wang, J.; Su, Y.; Wang, X.; Chen, J.; Zhao, Z.; Shen, M. The effect of partial substitution of Co in $LaMnO_3$ synthesized by sol–gel methods for NO oxidation. *Catal. Commun.* **2012**, *25*, 106–109. [CrossRef]
17. Taguchi, H.; Matsu-ura, S.-i.; Nagao, M.; Choso, T.; Tabata, K. Synthesis of $LaMnO_{3+\delta}$ by firing gels using citric acid. *J. Solid State Chem.* **1997**, *129*, 60–65. [CrossRef]
18. Taguchi, H.; Yamada, S.; Nagao, M.; Ichikawa, Y.; Tabata, K. Surface characterization of $LaCoO_3$ synthesized using citric acid. *Mater. Res. Bull.* **2002**, *37*, 69–74. [CrossRef]
19. Alifanti, M.; Auer, R.; Kirchnerova, J.; Thyrion, F.; Grange, P.; Delmon, B. Activity in methane combustion and sensitivity to sulfur poisoning of $La_{1-x}Ce_xMn_{1-y}Co_yO_3$ perovskite oxides. *Appl. Catal. B Environ.* **2003**, *41*, 71–81. [CrossRef]
20. Silva, C.R.B.; da Conceição, L.; Ribeiro, N.F.P.; Souza, M.M.V.M. Partial oxidation of methane over Ni–Co perovskite catalysts. *Catal. Commun.* **2011**, *12*, 665–668. [CrossRef]
21. Ghiasi, M.; Delgado-Jaime, M.U.; Malekzadeh, A.; Wang, R.-P.; Miedema, P.S.; Beye, M.; de Groot, F.M.F. Mn and Co charge and spin evolutions in $LaMn_{1-x}Co_xO_3$ nanoparticles. *J. Phys. Chem. C* **2016**, *120*, 8167–8174. [CrossRef]

22. Chen, J.; Shen, M.; Wang, X.; Qi, G.; Wang, J.; Li, W. The influence of nonstoichiometry on LaMnO$_3$ perovskite for catalytic NO oxidation. *Appl. Catal. B Environ.* **2013**, *134–135*, 251–257. [CrossRef]
23. Pecchi, G.; Campos, C.; Peña, O.; Cadus, L.E. Structural, magnetic and catalytic properties of perovskite-type mixed oxides LaMn$_{1-y}$Co$_y$O$_3$ (y = 0.0, 0.1, 0.3, 0.5, 0.7, 0.9, 1.0). *J. Mol. Catal. A Chem.* **2008**, *282*, 158–166. [CrossRef]
24. Rakshit, S.; Gopalakrishnan, P.S. Oxygen nonstoichiometry and its effect on the structure of LaNiO$_3$. *J. Solid State Chem.* **1994**, *110*, 28–31. [CrossRef]
25. Zhou, C.; Feng, Z.; Zhang, Y.; Hu, L.; Chen, R.; Shan, B.; Yin, H.; Wang, W.G.; Huang, A. Enhanced catalytic activity for NO oxidation over Ba doped LaCoO$_3$ catalyst. *RSC Adv.* **2015**, *5*, 28054–28059. [CrossRef]
26. Qi, G.; Li, W. Pt-free, LaMnO$_3$ based lean NO$_x$ trap catalysts. *Catal. Today* **2012**, *184*, 72–77. [CrossRef]
27. Batiot-Dupeyrat, C.; Valderrama, G.; Meneses, A.; Martinez, F.; Barrault, J.; Tatibouët, J.M. Pulse study of CO$_2$ reforming of methane over LaNiO$_3$. *Appl. Catal. A Gen.* **2003**, *248*, 143–151. [CrossRef]
28. Benard, S.; Retailleau, L.; Gaillard, F.; Vernoux, P.; Giroir-Fendler, A. Supported platinum catalysts for nitrogen oxide sensors. *Appl. Catal. B Environ.* **2005**, *55*, 11–21. [CrossRef]
29. Mulla, S.S.; Chen, N.; Cumaranatunge, L.; Blau, G.E.; Zemlyanov, D.Y.; Delgass, W.N.; Epling, W.S.; Ribeiro, F.H. Reaction of NO and O$_2$ to NO$_2$ on Pt: Kinetics and catalyst deactivation. *J. Catal.* **2006**, *241*, 389–399. [CrossRef]
30. Schmitz, P.J.; Kudla, R.J.; Drews, A.R.; Chen, A.E.; Lowe-Ma, C.K.; McCabe, R.W.; Schneider, W.F.; Goralski, C.T. NO oxidation over supported Pt: Impact of precursor, support, loading, and processing conditions evaluated via high throughput experimentation. *Appl. Catal. B Environ.* **2006**, *67*, 246–256. [CrossRef]
31. Wang, H.; Chen, H.; Wang, Y.; Lyu, Y.-K. Performance and mechanism comparison of manganese oxides at different valence states for catalytic oxidation of NO. *Chem. Eng. J.* **2019**, *361*, 1161–1172. [CrossRef]
32. Ivanova, S.; Senyshyn, A.; Zhecheva, E.; Tenchev, K.; Stoyanova, R.; Fuess, H. Crystal structure, microstructure and reducibility of LaNi$_x$Co$_{1-x}$O$_3$ and LaFe$_x$Co$_{1-x}$O$_3$ perovskites (0 <x ≤ 0.5). *J. Solid State Chem.* **2010**, *183*, 940–950.
33. Brunauer, S.; Emmett, P.H.; Teller, E. Adsorption of gases in multimolecular layers. *JACS* **1938**, *60*, 309–319. [CrossRef]
34. Scherrer, P. Bestimmung der größe und der inneren struktur von kolloidteilchen mittels röntgenstrahlen. *Nachr. Ges. Wiss. Gött. Math. Phys. Kl.* **1918**, *1918*, 98–100.
35. Tsakoumis, N.E.; Voronov, A.; Rønning, M.; Beek, W.v.; Borg, Ø.; Rytter, E.; Holmen, A. Fischer–tropsch synthesis: An XAS/XRPD combined in situ study from catalyst activation to deactivation. *J. Catal.* **2012**, *291*, 138–148. [CrossRef]
36. Fogler, H. *Elements of Chemical Reaction Engineering*; Prentice-Hall of India: New Delhi, India, 2006; p. 92.

 © 2019 by the authors. Licensee MDPI, Basel, Switzerland. This article is an open access article distributed under the terms and conditions of the Creative Commons Attribution (CC BY) license (http://creativecommons.org/licenses/by/4.0/).

Article

Reactive Grinding Synthesis of LaBO$_3$ (B: Mn, Fe) Perovskite; Properties for Toluene Total Oxidation

Bertrand Heidinger [1,2], Sébastien Royer [1,2], Houshang Alamdari [2], Jean-Marc Giraudon [1] and Jean-François Lamonier [1,2,*]

1. Univ. Lille, CNRS, Centrale Lille, ENSCL, Univ. Artois, UMR 8181–UCCS–Unité de Catalyse et Chimie du Solide, 59000 Lille, France
2. Department of Mining, Metallurgical and Materials Engineering, Laval University, Québec City, QC G1V 0A6, Canada
* Correspondence: jean-francois.lamonier@univ-lille.fr; Tel.: +33-320337733

Received: 2 July 2019; Accepted: 23 July 2019; Published: 25 July 2019

Abstract: LaBO$_3$ (B: Mn, Fe) perovskites were synthesized using a three-step reactive grinding process followed by a calcination at 400 °C for 3 h. The three successive steps are: (i) solid state synthesis (SSR); (ii) high-energy ball milling (HEBM); (iii) low-energy ball milling (LEBM) in wet conditions. The impact of each step of the synthesis on the material characteristics was deeply investigated using physico-chemical techniques (X-ray diffraction (XRD), N$_2$-physisorption, scanning electron microscopy (SEM), transmission electron microscopy (TEM), temperature-programmed reduction (H$_2$-TPR), X-ray photoelectron spectroscopy (XPS)) and the catalytic performances of the synthesized materials were evaluated for the toluene total oxidation reaction. Starting from single oxides, microcrystalline perovskite phase, exhibiting negligible surface areas, is obtained after the SSR step. The HEBM step leads to a drastic reduction of the mean crystal size down to ~20 nm, along with formation of dense aggregates. Due to this strong aggregation, surface area remains low, typically below 4 m$^2 \cdot$g^{-1}. In contrast, the second grinding step, namely LEBM, allows particle deagglomeration resulting in increasing the surface area up to 18.8 m$^2 \cdot$g^{-1} for LaFeO$_3$. Regardless of the perovskite composition, the performance toward toluene oxidation reaction increases at each step of the process: SSR < HEBM < LEBM.

Keywords: volatile organic compounds; catalytic oxidation; perovskite; reactive grinding; toluene

1. Introduction

Volatile organic compounds (VOCs) are responsible of important environmental and health issues such as greenhouse gas effect, tropospheric ozone accumulation or CMR (carcinogenic, mutagenic or reprotoxic) behavior on animals and human beings. Those chemicals are generated from both natural and anthropogenic sources, the latter being the most significant. Industry is a major contributor to the VOC emissions, with a huge consumption of chemicals, especially benzene, toluene and xylene (BTX) mainly used as precursors or organic solvents [1,2]. Toluene is, for example, a potential CMR and has consequent greenhouse power [3,4]. Professional exposure limits have been set up to protect workers with respectively 50 and 100 ppmv values for the long- and short-term exposition for the European Union [5]. Over the last two last decades, VOC emissions have been highly reduced to fit the stringent environmental regulations [2,6]. While chemical substitution has to be privileged, this alternative is difficult to adapt to industrial processes which are already well-optimized. Then, catalytic oxidation processes as a post-treatment solution, play a major role in VOC emissions control with attractive characteristics: limited energy consumption as well as complete and selective elimination of pollutants. Supported noble metal catalysts are active catalysts even at low temperature [7] but they generally suffer from a deactivation over time by poisoning [8,9]. Increasing the cost and rarefication of these resources limit their wide use [10]. Alternatives to noble metal catalysts are transition metal oxides, being of lower cost and some of them showing interesting catalytic performances in oxidation reactions [11,12]. Among them are perovskites-like mixed oxides, commonly described by the general formula ABO_3 where A is an alkaline, an alkaline earth or a rare-earth cation and B is a transition metal. With numerous possible A and B associations, properties of perovskite-like materials can be fine-tuned regarding the targeted application [13–15]. Properties of perovskites are closely related to their synthesis method [16–18]. Parameters being identified as crucial are specific surface area (SSA), crystal domain size and transition metal surface accessibility, all parameters having an effect on cation reducibility and then on catalytic activity. Among the different compositions, $LaMnO_{3.15}$ and $LaFeO_3$ are of particular interest. $LaMnO_{3.15}$ shows excellent catalytic performances related to Mn^{IV}/Mn^{III} mixed valence stabilized in the structure (charge neutrality being reached with cationic vacancies). Considering a lower reducibility of the iron cations in the crystal (leading to a lower activity than those of the Co- and Mn-counterparts), $LaFeO_3$ is far less studied, except for high temperature application due to its good stability [14].

As the important parameter that directly impacts the catalytic activity is the specific surface area. While solid state reaction route produces material with low specific surface area, not ideal in view of a catalytic application, solution-mediated synthesis routes give access to materials with better textural properties, but they can hardly be considered as a sustainable solution because of the consumption of solvent. On the other hand, reactive grinding (RG) is a common approach used in metallurgy that shows attractive features such as flexibility, low temperature, atmospheric pressure, no use of solvents (or in a small amount) [19,20]. Reactive grinding has then already been proved to be efficient to produce catalysts such as MnO_2, several hexaaluminates and perovskites [21–24]. In this work, nanocrystalline $LaMnO_3$ and $LaFeO_3$ perovskites-type mixed oxides, exhibiting high specific surface areas, are obtained by a three-step reactive grinding synthesis. The selected synthesis sequence consists in: (1) a solid-state reaction step, starting with selected single oxides to obtain the perovskite phase, (2) the structural modification (crystal size decrease) with high-energy ball milling (HEBM) step, and finally, (3) a low-energy ball milling (LEBM) for surface area development. Solids are characterized at each step of the synthesis, and catalytic performances as well as stability behavior of $LaMnO_{3.15}$ and $LaFeO_3$, are reported for the toluene total oxidation reaction.

2. Results

2.1. Physico-Chemical Characterization

2.1.1. Structural Properties Evolution upon Grinding

X-ray diffractograms obtained for $LaMnO_{3.15}$ and $LaFeO_3$ materials, after each step of the synthesis, are shown in Figure 1. In the case of $LaMnO_{3.15}$, after solid state synthesis step (LaMn_SSR), intense and narrow diffraction peaks that match with the $LaMnO_{3.15}$ reference pattern (PDF#50-0298) of rhombohedral structure are observed. In addition to the main perovskite phase, weak remaining peaks (2θ = 26.2°, 27.9°, 29.2°, 29.9°, 36.1°, 39.6°, 46.1°, 55.5 or 59.9°) are also observed. These reflections are associated to unreacted oxide precursors and derived phases: La_2O_3 (PDF#74-2430), $La(OH)_3$ (PDF#36-1481) and Mn_3O_4 (PDF#18-0803). However, these reflections, weak in intensity, suggest that most of the material consists in crystalline perovskite. After the high energy ball milling step, diffractogram recorded for LaMn_HEBM shows the comparable reflections, nonetheless the peaks are less intense and broadened. In addition, the peaks associated to the unreacted phases disappeared. This kind of evolution was already reported in previous works [22,25], and corresponds to a significant decrease of the coherent crystal domain. After the second step of grinding, namely low energy ball milling (LEBM performed in wet conditions), the obtained diffractograms are very similar to that obtained for the LaMn_HEBM sample. No significant evolution, neither for the phase detected nor the intensity/width of the reflection, can be observed. Comparable evolution of the Fe-containing perovskite diffractograms with the synthesis step is observed. Initially, LaFe_SSR displays $LaFeO_3$

perovskite-type phase (PDF#37-1493, Orthorhombic crystal structure) in addition to La(OH)$_3$ (peaks at 2θ = 27.4°, 28°, PDF#36-1481) and Fe$_2$O$_3$ (peaks at 2θ = 33.2°, 35.6°, PDF#33-1481) impurity phases. After the HEBM step, perovskite reflection broadening is observed, in addition to the disappearance of reflections ascribed to unreacted phases, and finally, after LEBM step, a comparable diffractogram to that obtained previously is obtained.

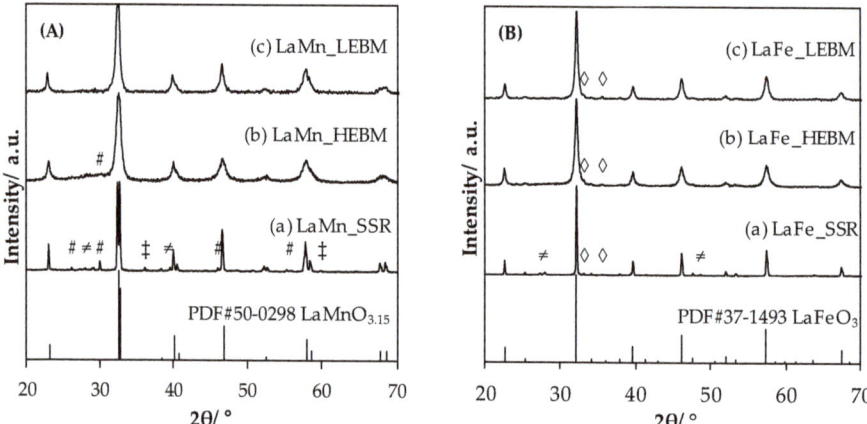

Figure 1. Diffractograms obtained for LaMnO$_{3.15}$ (**A**) and LaFeO$_3$ (**B**) after each synthesis step. SSR: solid state reaction, HEBM: high-energy ball milling, LEBM: low-energy ball milling. Bottom of the figure: vertical bars are for cited JCPDS reference. #, La$_2$O$_3$; ‡, Mn$_3$O$_4$; ≠, La(OH)$_3$; ◊, Fe$_2$O$_3$.

Crystal domain sizes were estimated using the Scherrer equation after correction for instrumental broadening. The obtained values are summarized in Table 1. Starting with micrometric crystals for the SSR-derived materials the crystal size drops down to a value in the size range of 14 to 22 nm depending on the perovskite composition and milling step. In that respect, the HEBM step induced a significant decrease of the crystal size attesting in that way that the energy transferred by the grinding induces a particle breakage. On the contrary, no significant modification of the crystal domain size could be observed after LEBM showing that this low-energetic wet milling process precludes additional decrease of the crystals. Interestingly, the B-metal nature did not have a significantly impact on the final crystal domain size which ranged in the 15–20 nm interval in line with a previous work reported in the case of Mn-containing perovskites [22] and slightly above the values reported by Gashdi et al. [25] for La$_{1-x}$Ce$_x$CoO$_3$ (11–13 nm). However, in this last reference, the authors did not calcine their materials for characterization and application purpose that can explain the slightly lower reported values.

Table 1. Structural, textural properties and elemental composition obtained for LaMnO$_{3.15}$ and LaFeO$_3$ after each synthesis step.

Sample	XRD Phase	D$_{cryst}$ (nm)	SSA [2] (m^2·g^{-1})	Fe cont. [3] (at.%)
LaMn_SSR	P, La$_2$O$_3$, La(OH)$_3$, Mn$_3$O$_4$	>500 [1]	1.0	0.0
LaMn_HEBM	P, La$_2$O$_3$	14	3.5	0.7
LaMn_LEBM	P	16	9.9	2.9
LaFe_SSR	P, La(OH)$_3$, Fe$_2$O$_3$	>500 [1]	1.6	-
LaFe_HEBM	P, Fe$_2$O$_3$	17	3.4	-
LaFe_LEBM	P, Fe$_2$O$_3$	22	18.8	-

P: perovskite, TM: transition metal; [1] estimated by SEM; [2] SSA: specific surface area; [3] measured by X-ray fluorescence spectroscopy.

2.1.2. Textural Properties Evolution upon Grinding

N_2-adsorption/desorption isotherms over LaFe based samples are presented in Figure 2. The curves exhibit the same pattern of a type II isotherm, regardless of the material composition and this isotherm shape is characteristic of materials displaying no significant porosity and the sharp N_2 adsorption occurring at P/P_0 above 0.8 is characteristic of external aggregate/particle porosity [26]. The amount of adsorbed N_2 for the SSR material (shown for $LaFeO_3$, Figure 2) is very low, demonstrating a limited pore volume. After the high-energy ball milling step, an increase of adsorbed N_2 was observed but total volume adsorbed remained relatively low. The impact of the LEBM step on the total N_2 volume adsorbed was more noticeable, as observed in Figure 2, showing that this step will impact the global porosity of the material, even if it was previously observed to have a very limited effect on the crystal domain size (Figure 1 and Table 1). As expected, the SSR materials, obtained at high crystallization temperature, displayed very limited surface areas (Table 1, ranging from 1.0 to 1.6 $m^2 \cdot g^{-1}$). This low surface area was directly related to the large crystal domain size displayed by the material (Table 1). After the HEBM step, leading to a significant decrease of the crystal domain size, the surface area slightly increased to 3.5 $m^2 \cdot g^{-1}$ ± 0.1 $m^2 \cdot g^{-1}$ for LaTM_HEBM, samples. Interestingly, the significant decrease of the crystal size did not promote the SSA of the samples as expected for an increase of the external surface due to non-interacting low size crystallites. Indeed, taking into account a $LaFeO_3$ having a mean crystallite size of spherical-shape amounting to 17 nm and a density of 6.65 a theoretical SSA of 53.1 $m^2 \cdot g^{-1}$ was expected. The large discrepancy between the experimental surface area (3.4 $m^2 \cdot g^{-1}$ of accessible surface to N_2) and the theoretical one indicated the formation of nanoparticle agglomerates displaying very limited porosity. After LEBM, a significant increase of the SSA was observed despite the absence of crystal domain size evolution during the milling. Then, LaFe_LEBM displayed a surface area of 18.8 $m^2 \cdot g^{-1}$ five time larger than the parent one. Such evolution demonstrated that the LEBM is acting mostly on the morphology of the aggregates present in the material. Two phenomena can be the origin of the SSA increase: (i) the decrease of the agglomerate size (deagglomeration process), (ii) formation/stabilization of a new porosity in the formed agglomerates [21,27]. However, it is very difficult to conclude on the impact of each phenomenon on the total surface increase, based only on N_2 physisorption results.

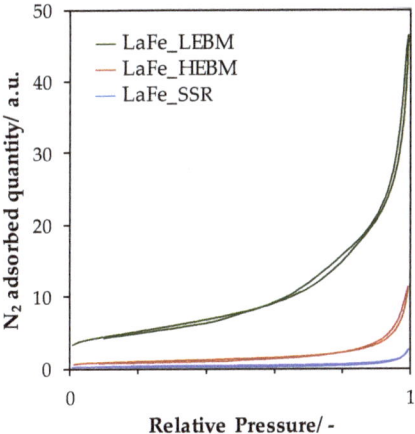

Figure 2. N_2-Physisorption isotherms recorded for $LaFeO_3$ samples after each step of synthesis.

2.1.3. Morphology Evolution upon Grinding

Evolution of the morphology of the LaTM perovskite after each synthesis step was observed by scanning electron microscopy (SEM) and transmission electron microscopy (TEM). It should be noted that similar observations hold whatever TM is concerned. As an example, the corresponding

SEM images for LaMn based samples are given in Figure 3a–c. The SSR sample, synthesized at high temperature (1100 °C), exhibited large hexagonal particles of different sizes, most of them being above 500 nm large and agglomerated. After HEBM, the morphology of the material consisted in small particles aggregated into large objects of about one micron in size. Then, the HEBM step allowed us to reduce the elementary crystallite size, which was consistent with the observed evolution from X-ray diffraction (XRD) analysis.

Figure 3. Scanning electron microscope (SEM) images obtained for LaMnO$_{3.15}$ (left) and transmission electron microscope (TEM) images obtained for LaFeO$_3$ (right), after each synthesis steps. (**a,d**) SSR: solid state reaction, (**b,e**) HEBM: high-energy ball milling, (**c,f**) LEBM: low-energy ball milling.

TEM analysis, presented in Figure 3d–f for LaFe-based samples, allowed us to observe the elementary particles (Figure 3e) which ranged from 10 to 40 nm. These sizes were in accordance with the XRD results, while the low surface area determined after the HEBM can be explained by the formation of dense, poorly porous, aggregates. When the material is subjected to LEBM, no significant modification of the large-scale morphology could be observed by SEM. Indeed, only aggregates of small nanocrystals were detected (Figure 3c). However, observation of the sample by TEM evidenced the formation of lower size aggregate/elementary particles despite the elementary particles remaining of comparable size than for the HEBM-derived material. As already concluded from XRD analysis, the LEBM step did not allow crystal size (as confirmed by TEM analysis, Figure 3f) to decrease, but seemed to allow the deagglomeration of the particles, leading in fine to the production of a material with higher SSA than after the HEBM step (Table 1). The Scheme 1 summarizes the role of the individual process steps, with the main characteristics of the perovskite obtained.

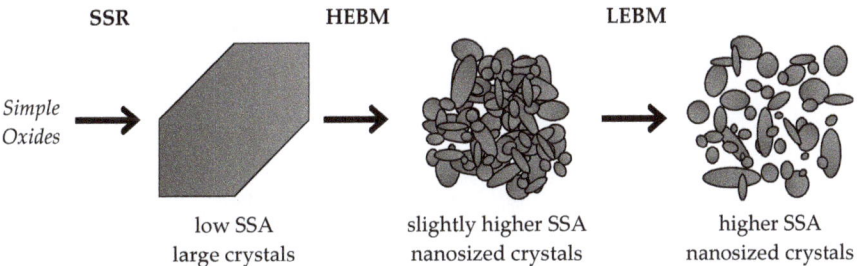

Scheme 1. Simplified view of the material evolution with the synthesis steps of the reactive grinding process. SSR: solid state reaction; HEBM: high energy ball milling; LEBM: low energy ball milling.

2.1.4. Evolution of Transition Metal Reducibility upon Grinding

Iron contamination is classically observed as a direct consequence of the grinding process [17,21,28] and has been estimated for LaMn samples (Table 1). Values staying below 3.0 at.% are reasonable even if an impact on the catalytic properties cannot be excluded, as already observed for MnO_x [21].

Temperature-programmed reduction (H_2-TPR) profiles of $LaMnO_{3.15}$ (A) and $LaFeO_3$ (B) materials are shown in Figure 4. Quantification of consumed hydrogen and temperatures at maximum hydrogen consumption are listed in Table 2.

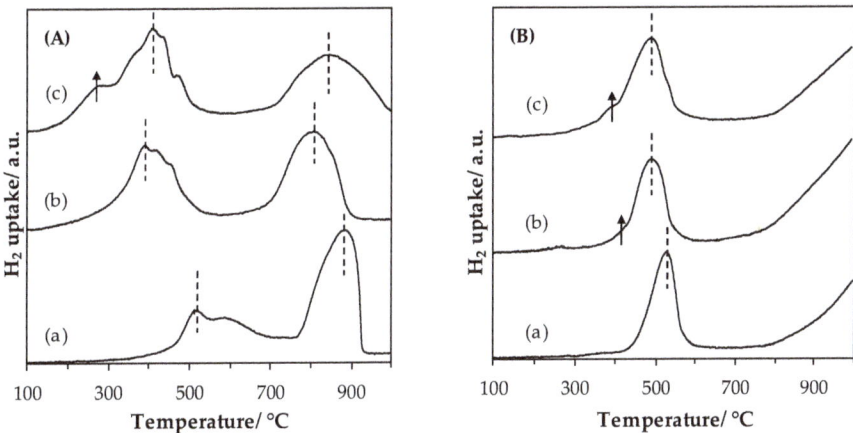

Figure 4. Temperature-programmed reduction (H_2-TPR) profiles of $LaMnO_{3.15}$ (**A**) and $LaFeO_3$ (**B**) samples after the (a) SSR, solid state reaction; (b) HEBM, high-energy ball milling; (c) LEBM, low-energy ball milling, steps.

Table 2. H_2-TPR results for $LaMnO_{3.15}$ and $LaFeO_3$ samples.

Sample	$T_{1,max}$ (°C)	$T_{2,max}$ (°C)	H_2 Uptake [1] (mmol(H_2)·g^{-1})	TM AOS [2]
LaMn_SSR	512	882	1.04 (2.63)	3.3
LaMn_HEBM	390	806	1.54 (3.02)	3.4
LaMn_LEBM	404	845	2.09 (3.14)	3.4
LaFe_SSR	523	-	0.89	2.4
LaFe_HEBM	490	-	0.92	2.5
LaFe_LEBM	489	-	1.18	2.6

TM: transition metal; [1] H_2 uptake from the low temperature (LT) consumption peak domain, total uptake being into brackets; [2], calculated for $LaMnO_{3.15}$ from the total H_2 uptake assuming a complete reduction up to Mn(+II), and calculated for $LaFeO_3$ from the LT H_2 uptake and assuming a reduction to Fe(+II) during this LT process, assuming bulk La/MT ratio of 1.

Reduction of $LaMnO_{3.15}$: the H_2 consumption profile displays 2 main peaks for all the LaMn based catalysts. Overall, the Mn(+IV) and Mn(+III) entities initially present in the perovskite structure were reduced to Mn(+II). According to the literature, the two H_2 consumption peaks can be ascribed to the consecutive Mn(+IV) → Mn(+III) and the Mn(+III) → Mn(+II) reductions [29,30]. Regarding the LaMn_SSR sample, these peaks were located at ~510 °C (shoulder: ~590 °C) and ~880 °C. Comparatively, for the LaMn_HEBM sample there was a lowering in terms of initial H_2 consumption temperature as well as of the position of the peak maximum indicating an enhancement in the reducibility of the sample. Such improvements in reducibility have been already observed for the reduction of $LaCoO_3$ [17] and have been interpreted in terms of a decrease of the crystal domain size. Finally, the reduction profile of LaMn_LEBM is globally comparable to that for LaMn_HEBM. However, it was now found at low temperature a new H_2-consumption shoulder. This observation indicates that the LEBM plays a beneficial role in promoting the reducibility of the sample through particle deagglomeration. The Mn AOS estimated from the global H_2 uptake gave a value of 3.4 (±0.1) in line with a $LaMnO_{3.15}$ phase detected by XRD.

Reduction of $LaFeO_3$: as for the Mn-based materials, two reduction steps were observed on the reduction profile of Fe-based samples. The hydrogen consumption profile registered over the LaFe_SSR sample shows one first reduction peak, with a maximum consumption at 520 °C. A second hydrogen consumption was observed to start at T > 750 °C, and is not achieved at the end of the experiment, i.e., at 1000 °C. According to the literature [31], and knowing that iron is essentially at the +III oxidation state in $LaFeO_3$, the two successive steps should be described as:

- Complete reduction of Fe(+III) to Fe(+II), with a possible additional hydrogen consumption of surface Fe(+II) reduction,
- Reduction of remaining Fe(+II) to Fe(0), and obtaining of La_2O_3 + Fe(0) individual phases.

Quantification of hydrogen consumed in the first reduction peak, for LaFe_SSR allowed us to calculate an AOS of 2.4 (considering that the first step is only related to the Fe(+III) to Fe(+II) reaction), that is consistent with results available in the literature [31]. As observed for the $LaMnO_{3.15}$ series, the HEBM step led to a shift of the hydrogen consumption step toward the low temperatures by 30 °C with a small additional consumption visible at 410 °C. Quantification led to a comparable value of AOS (Table 2). Finally, the LEBM step does not induce a significant change in hydrogen consumed and hydrogen consumption position. The only minor modification concerns the low temperature contribution, and the increase in intensity of the consumption located at 410 °C. However, global consumption always leads to an AOS of 2.5 (±0.1), comparable to the value obtained for the SSR and HEBM materials. The consumption peak at 410 °C can originate from the reduction of surface FeOx clusters, since Faye et al. reported a comparable temperature of reduction of Fe_2O_3 highly dispersed on the surface of $LaFeO_3$ (405 °C for the first consumption [31]).

2.1.5. Surface Properties of Materials

X-ray photoelectron spectroscopy (XPS) quantification results are shown in Table 3. The quantification is based on La 3d, O 1s and Mn 2p or Fe 2p peak areas. The La atomic percentages are higher than stoichiometry. The lanthanum surface enrichment is often reported for La-based perovskites [32]. For LaMn_SSR sample, this enrichment can be explained by the presence of La(OH)$_3$ at the outermost surface. With the disappearance of La(OH)$_3$ in the course of ball milling and the surface contamination by Fe, the La enrichment became less pronounced for LaMn_HEBM and LaMn_LEBM samples. An opposite trend was observed over the LaFeO$_3$ samples as a slight La enrichment of the surface was observed after the HEBM and LEBM steps, a phenomenon that cannot be explained at this stage considering that TPR experiment, and especially the low contribution reduction at 410 °C, suggesting the formation of surface FeOx cluster species that should have contributed to a decrease in La/Fe surface ratio.

Table 3. Quantification from X-ray photoelectron spectroscopy (XPS) analysis for LaMnO$_{3.15}$ and LaFeO$_3$ samples.

Sample	La (at.%)	Mn (at.%)	Fe (at.%)	La/(Fe+Mn) (-)	O$_{total}$ (at.%)	O$_I$ (at.%)	O$_{II}$ (at.%)	O$_{III}$ (at.%)	AOS [1]
LaMn_SSR	19.2	9.9	-	2.01	70.9	20.9	44.9	5.2	3.2
LaMn_HEBM	16.4	11.3	n.q.	1.45	72.3	29.4	32.3	10.6	3.8
LaMn_LEBM	16.3	11.5	1.4	1.26	70.7	33.0	32.8	4.9	3.7
LaFe_SSR	14.4	-	9.1	1.58	76.5	30.7	33.2	12.7	-
LaFe_HEBM	19.9	-	10.0	1.99	70.1	32.0	31.7	6.4	-
LaFe_LEBM	19.0	-	11.1	1.71	69.9	33.2	29.5	7.2	-

[1] issued from ΔE(Mn 3s) values.

The superposition of the O 1s spectra for the three LaMnO$_{3.15}$ samples is shown in Figure 5. Satisfactory peak fitting can be achieved with three components in the three samples. The component O$_I$ at low BE (529.2 eV) is assigned to bulk O^{2-} species [33]. At intermediate BE, the large O$_{II}$ component, centered at 531.3 eV, is ascribed to several species such as OH$^-$, CO$_3^{2-}$, O$_2^{2-}$ and/or O$^-$ species [34]. The presence of carbonate species is confirmed on the C 1s core level spectra, with a signal located at 289.0 eV. The O$_{III}$ component, at higher BE (533.3 eV), originates from adsorbed water [35]. The SSR sample exhibits an intense O$_{II}$ component, which can be related to the presence of La(OH)$_3$ (XRD) and carbonate surface species. The contribution of this component to the O 1s signal significantly decreases after the HEBM process and remains rather constant after the LEBM process, in agreement with the disappearance of the La(OH)$_3$ phase after the ball milling processes. Figure A1 (Appendix A) shows the La 3d region for LaMnO$_{3.15}$ samples, the La 3d spectrum being split into a 3d$_{5/2}$ and a 3d$_{3/2}$ lines due to the spin-orbit interaction. The magnitude of the multiplet splitting can be useful for the chemical assessment. While the energy difference between the main peak and its satellite (ΔE(La3d)) is around 3.9 for La(OH)$_3$, the one for La$_2$O$_3$ is higher (4.6 eV) [36]. Therefore the increase in ΔE(La3d) value observed after the HEBM process suggests a lower La(OH)$_3$ contribution to the La3d signal, in agreement with the XRD results and O 1s spectra analysis. Mn 2p spectra (Figure A1) show two main peaks corresponding to the spin–orbit split of 2p$_{3/2}$ and 2p$_{1/2}$ levels, while the weak signal at lower BE from the main peak is assigned to the satellite of the 2p$_{1/2}$ peak. The 2p$_{3/2}$ peak satellite is not noticeable because it overlaps with the 2p$_{1/2}$ peak. Mn 2p$_{3/2}$ peak has its maximum at BE of 641.6 eV for the three LaMnO$_{3.15}$ samples. This BE value is intermediate between those recorded for Mn$_2$O$_3$ and for MnO$_2$ [37], confirming the presence of a mixture of Mn^{3+} and Mn^{4+} species in the LaMnO$_{3.15}$ samples. In order to estimate the proportion of Mn^{3+} and Mn^{4+} species, the Mn 3s core level has been studied. The superposition of Mn 3s spectra is shown in Figure 5. Two peaks, originating from the coupling of non-ionized 3s electron with 3d valence-band electrons [38], are distinguished. From the energy difference between the two peaks, ΔE(Mn 3s), it is possible to estimate the Mn AOS [39]. A significant decrease in ΔE(Mn 3s) value was observed from LaMn_SSR to LaMn_HEBM sample, and then remains stable for LaMn_LEBM sample (Figure 5). Therefore, the HEBM process resulted in a pronounced increase in Mn AOS on the material surface from 3.2 to 3.8, while the LEBM process does not induce significant additional change (Table 3). The surface AOS (obtained by XPS, Table 3) is identical to the bulk AOS (obtained by TPR, Table 2) for the LaMn_SSR material. However, while the bulk AOS was not affected during the HEBM and LEBM steps, it is evident from the XPS results that these steps result in an oxidation of the surface Mn ions.

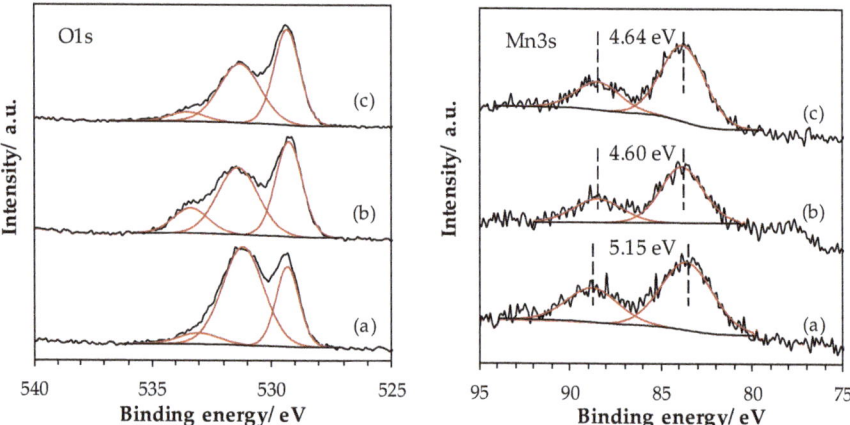

Figure 5. High-resolution spectra of O 1s and Mn 3s core level spectra for LaMnO$_{3.15}$ samples: (a) SSR, solid state reaction; (b) HEBM, high-energy ball milling; (c) LEBM, low-energy ball milling.

XPS results obtained for the LaFeO$_3$ samples are presented in Figure 6. The superposition of the O 1s spectra is shown in Figure 6. As previously observed for Mn-based samples, the same three components can be extracted from the O 1s signal. However, the component contributions to the O 1s signal are not significantly impacted by the ball milling process, since the O$_I$ to O$_{II}$ component ratio is similar regardless the synthesis step. Figure A2 shows the La 3d region obtained for the LaFe samples. ΔE(La 3d) values remains constant for the three samples: the multiplet splitting amplitude of ~4.2 eV confirms the oxide form. Figure 6 shows the curve fitted Fe 2p spectra. Indeed, the similarly to Mn 2p spectra, Fe 2p signal presents multiplet structures which can be fitted in order to resolve the surface iron state. The Fe 2p spectral fitting parameters of Fe$_2$O$_3$ compound proposed by Biesinger et al. [37], i.e., binding energy, percentage of total area, full width at half maximum and spectral component separation, have been used to simulate the Fe 2p signal. The good concordance between the fitted integration and the Fe 2p experimental envelope suggests that iron species are mainly in +III oxidation state, as it is often described in the literature.

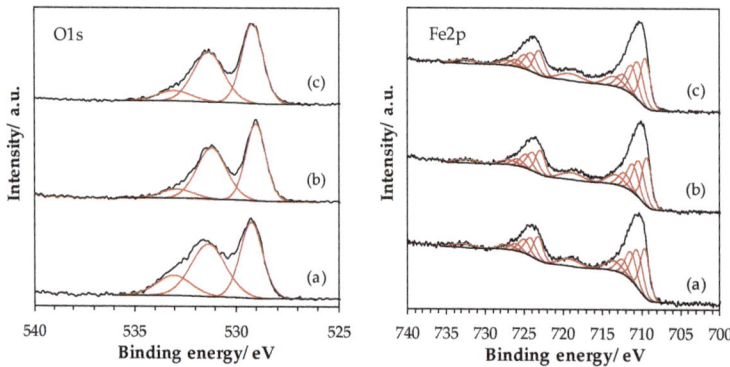

Figure 6. High-resolution spectra of O 1s and Fe 2p core level spectra for LaFeO$_3$ samples: (a) SSR, solid state reaction; (b) HEBM, high energy ball milling; (c) LEBM, low energy ball milling.

2.2. Catalytic Performances

2.2.1. Catalysts Activity

Catalytic performances of materials were evaluated in the toluene total oxidation reaction. Conversion versus reaction temperature curves are shown in Figure 7. As first remark, reactions performed with different LaFe samples lead to the production of only CO_2 and water as products, while benzene traces (not quantifiable) appear as a byproduct during reactions over LaMn samples, for high conversion value ($X > 50\%$). Consequently, conversions plotted in Figure 7 refer to selective conversion into CO_2.

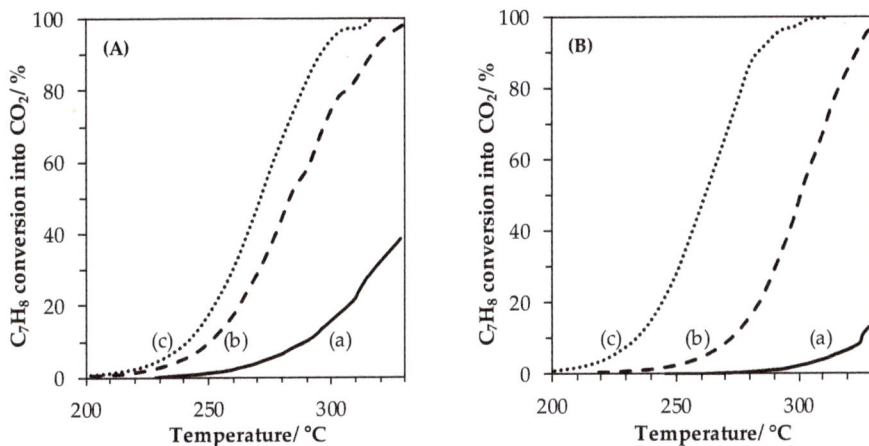

Figure 7. Toluene conversion curves obtained for $LaMnO_{3.15}$ (**A**) and $LaFeO_3$ (**B**) samples: (a) SSR, solid state reaction; (b) HEBM, high-energy ball milling; (c) LEBM, low-energy ball milling. Conditions: 200 mg of catalyst, 100 mL·min^{-1} of 1000 ppmv C_7H_8 in synthetic air (20% O_2 in N_2).

LaMn_SSR and LaFe_SSR (straight lines) show low performances for the toluene oxidation reaction with 39% and 13% toluene converted in CO_2 at T = 330 °C, respectively. After the high-energy ball milling process, far better performances were obtained with a shift of the light-off curves toward the lower temperature by −39 °C (LaMn_HEBM) and −52 °C (LaFe_HEBM). Samples from the second step of grinding at low energy showed even better performances with additional shifts toward the lower temperature by −12 °C for LaMn_LEBM, and by −40 °C for LaFe_LEBM. T_{10}, T_{50} and T_{90} values, reflecting the catalytic performances of each materials are given in Table 4. The following ranking of activity was obtained:

Table 4. T_{50} values and kinetics data obtain from light-off curves of LaMn and LaFe samples.

Sample	T_{10} (°C)	T_{50} (°C)	T_{90} (°C)	Activity [1] (mmol·s^{-1}·m^{-2})	Ea [2] (kJ·mol^{-1})	A_0 [3] (s^{-1})
LaMn_SSR	290	/	/	/	126	3.2×10^{12}
LaMn_HEBM	251	283	317	8.68×10^{-6}	146	2.6×10^{13}
LaMn_LEBM	242	271	295	6.41×10^{-6}	140	5.0×10^{13}
LaFe_SSR	325	/	/	/	164	3.6×10^{11}
LaFe_HEBM	273	301	323	2.44×10^{-6}	155	6.5×10^{12}
LaFe_LEBM	234	261	283	5.39×10^{-6}	137	8.7×10^{13}

[1] determined at T = 250 °C; [2] determined in the range $X(\%) < 20\%$; [3] extrapolated using the Ea average value.

LaFe_LEBM > LaMn_LEBM > LaMn_HEBM > LaFe_HEBM > LaMn_SSR > LaFe_SSR

Then, and as easily observed in Figure 7, the most active materials are LEBM-derived solids. It is interesting to note that LaFe_HEBM is little more active than LaMn_HEBM, while in the literature, a reverse order is generally reported even at low temperature (CO [11,15]) and at high temperature (CH$_4$ [15,40]). However, when analyzing the normalized activities (per surface unit), the activity ranking becomes:

LaMn_HEBM > LaMn_LEBM > LaFe_LEBM > LaFe_HEBM >> LaMn_SSR > LaFe_SSR

This result clearly demonstrates that the Mn-containing formulation are little more active than the Fe-containing ones, in line with the more important reducibility of manganese as determined by TPR (Figure 4) and demonstrating a lower temperature of activation for the Mn(+IV)/Mn(+III) redox couple than for the Fe(+III)/Fe(+II) redox couple.

Figure 8 shows the Arrhenius plots drawn from conversion curves at X < 20%, assuming a first order toward toluene concentration and a zero order toward oxygen. Activation energies, estimated from Arrhenius plots slopes, are reported in Table 4. Calculated activation energies for the toluene oxidation over LaMn samples show values comprised between 126 and 146 kJ·mol^{-1}. Calculated activation energies for the reaction over LaFe samples oscillated from 137 to 164 kJ·mol^{-1} respectively. Considering the precision of the measure, i.e., light off curve measurement, whole materials display comparable E$_a$, at an average value of 145 kJ·mol^{-1} suggesting a comparable oxidation mechanism for LaFe- and LaMn-containing formulation and whatever the synthesis step (SSR, HEBM, LEBM). Similar activation energy values had been reported for La$_{1-x}$Ca$_x$BO$_3$ (B = Fe, Ni) systems by Pecchi et al. [41].

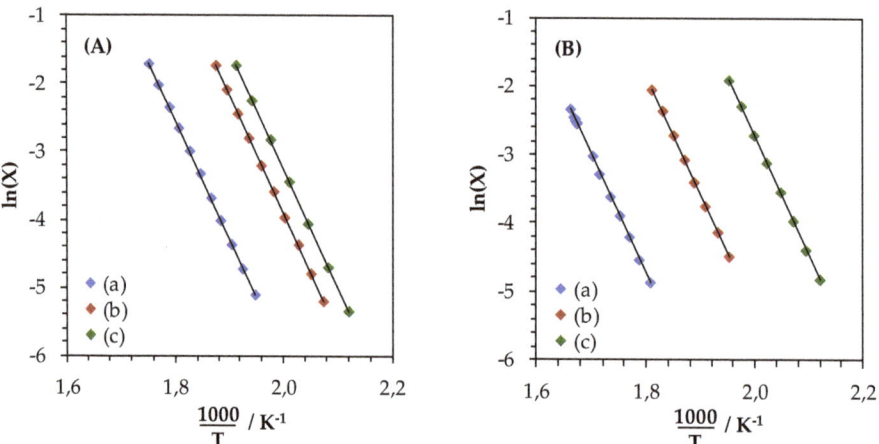

Figure 8. Arrhenius plots obtained at X < 20% for the toluene oxidation reaction over LaMnO$_{3.15}$ (**A**) and LaFeO$_3$ (**B**) samples: (a) SSR, solid state reaction; (b) HEBM, high-energy ball milling; (c) LEBM, low-energy ball milling.

Assuming an average E$_a$ of 145 kJ·mol^{-1}, corrected pre-exponential factor (A$_{0\,cor}$) is re-calculated, and obtained values are listed in Table 4. A$_{0\,cor}$ factors, obtained at constant E$_a$ for all materials, reflect the evolution in active sites on the catalysts, and when plotted as a function of the surface area of the solid (Figure 9), this allows us to determine the active site surface density. From Figure 9, it seems evident that a roughly linear correlation was obtained between A$_{0\,cor}$ and materials surface area. Consequently, the active sites in materials evolves linearly with the surface area. Nonetheless, LaMn samples showed a slightly higher active site surface density than their LaFe equivalent.

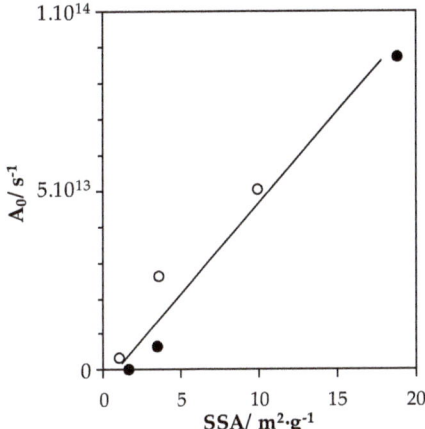

Figure 9. Evolution of corrected pre-exponential factors with catalyst surface areas. Open symbols: LaMn-compositions; Full symbols: LaFe-compositions.

2.2.2. Stability Tests

The toluene conversion into CO_2 evolution as a function of time are presented in Figure 10 for the HEBM and LEBM catalysts (T = 285 °C; 70 h). Both LaMn based catalysts show a rapid deactivation in the five first hours before to slightly linearly deactivate with time. The activity coefficient a_{285} (see experimental part) are rather similar for both catalysts amounting to 0.66 and 0.64 for LaTM_HEBM and LaTM_LEBM catalysts, respectively. This likely indicates that the LEBM treatment has no effect on the stability of the catalyst on stream. By opposition, a good stability for the LaFe-based catalysts was found over time in terms of toluene conversion into CO_2. It should be noted that, for all catalysts, it is found neither phase transformations nor crystallite size increase by XRD. Furthermore, the SSA keeps unchanged after the stability test considering the margin of uncertainties.

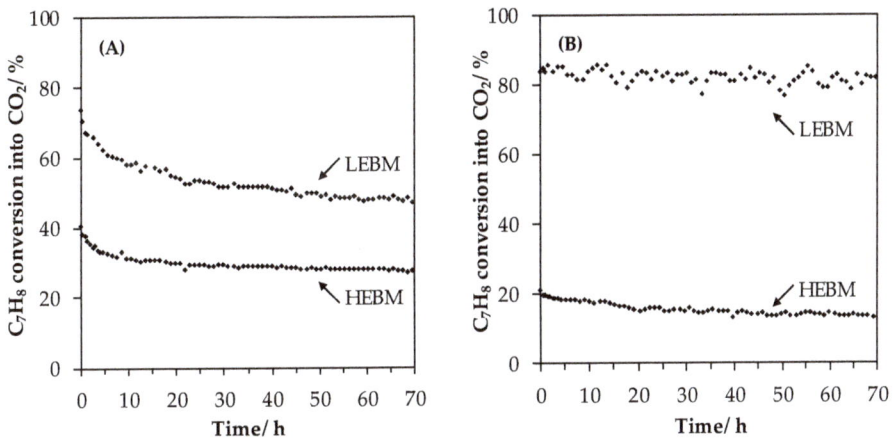

Figure 10. Stability experiments of $LaMnO_{3.15}$ (**A**) and $LaFeO_3$ (**B**) samples. Conditions are: 200 mg of catalyst exposed to 1000 ppmv C_7H_8 in 100 mL·min^{-1} of synthetic air (20% O_2 in N_2) for 70 h at a constant temperature of 285 °C. HEBM, high-energy ball milling; LEBM, low-energy ball milling.

Behar et al. have shown that the total oxidation of toluene can be easily described by a Mars–van Krevelen model when Cu-Mn mixed oxide is used as catalyst [42]. In this model the toluene is oxidized by the catalyst and not directly by the gaseous oxygen, and the role of gaseous oxygen was in restoring and maintaining the oxidized state of the catalyst. The oxidized state of the LaMn_LEBM after the stability experiment was studied by means of H_2-TPR and XPS analyses. Figure 11A shows the comparison between the two H_2-TPR profiles obtained on the fresh and used LaMn_LEBM catalysts. It can be clearly seen that, for the used LaMn_LEBM sample, the shoulder at low temperature (280 °C) disappeared, suggesting that some Mn species in a high oxidation state (Mn(+IV)) are reduced during the stability test. Similarly, the analysis of the Mn 3s photopic (Figure 11B) shows an increase in the peak splitting ΔE(Mn 3s) value, from 4.64 eV for fresh LaMn_LEBM to 4.91 eV for the used one, which demonstrates a decrease in the Mn AOS value (from 3.7 to 3.4) after the stability experiment. Therefore, the deactivation observed for LaMn_LEBM catalyst could be related to the decrease in Mn AOS, the step of reoxidizing the reduced catalyst with gaseous oxygen being most likely the rate-determining step.

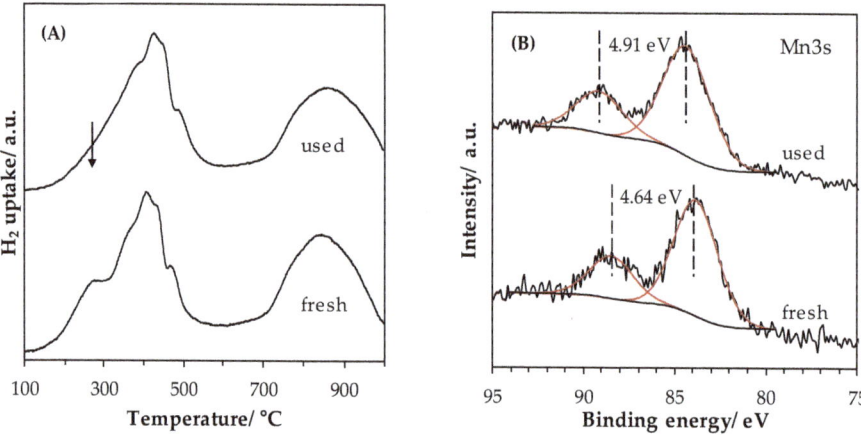

Figure 11. Comparison of (**A**) H_2-TPR profiles and (**B**) XPS high resolution spectra of Mn3s obtained for the fresh and used LaMn_LEBM catalysts.

3. Materials and Methods

3.1. Synthesis

Perovskite-type mixed metal oxides LaMO$_3$ (M = Mn or Fe) were synthesized by a three-step reactive grinding process:

- STEP 1, Solid state reaction—SSR: oxide precursors (La$_2$O$_3$ and Mn$_2$O$_3$ or Fe$_3$O$_4$) were homogeneously mixed in such a way as the molar La:B be equal to 1.0 and calcined for 4 h at 1100 °C under static air atmosphere to obtain the perovskite phase (confirmed by XRD analysis). Materials produced at this step were LaMn_SSR and LaFe_SSR.
- STEP 2, High-energy ball milling—HEBM: a first high energy milling step was performed for 90 min (SPEX 8000D grinder with stainless steel equipment and balls (ø 1 × 11 mm, 2 × 12.5 mm), under static air atmosphere) at a revolution frequency of 17.5 Hz (1060 cycle·min^{-1}).
- STEP 3, Low-energy ball milling—LEBM: a second low energy ball milling was performed for 120 min (Union Process' Svegvari attritor with stainless steel equipment and balls (ø 5 mm), with addition of a small amount of water (0.4 mL per g of material)) at a rotation speed of 450 rpm. The muddy sample was recovered using 500 mL of water and dried at 150 °C overnight.

Materials taken from step 2 and 3 are calcined for 3 h at 400 °C prior to characterization and catalytic performance evaluation, and are respectively named LaM_HEBM and LaM_LEBM.

3.2. Physico-Chemical Characterizations

The physicochemical properties of the materials were evaluated after each steps of the synthesis. Diffractograms are recorded on a Bruker D8 apparatus, using the CuKα radiation (λ = 1.54059 Å). Data were collected for 2θ between 10° and 80°, with an increment of 0.05° and an integration time of 1.0 s at each step. Diffractograms were indexed using references PDF database. Crystallite mean sizes are calculated using the Scherrer equation: $D_{cryst} = (k \cdot \lambda)/(\beta \cdot cos\theta)$, where k and β are respectively the shape factor (~0.9) and the corrected full width at half maximum. N_2-physisorption experiments were collected on a Micromeritics Tristar II porosity instrument. A known mass of catalyst was degassed at 150 °C under vacuum for 6 h. Isotherms were registered at a temperature of −196 °C and SSA were determined from the adsorption branch in the 0.05–0.30 P/P_0 range with the Brunauer–Emmett–Teller (BET) equation. SEM images were acquired on a JOEL JSM 5300 microscope equipped with EDXS module, allowing elemental quantification. Samples were previously put on a carbon tape and coated with graphitic carbon to reduce charging effect. TEM images were registered on a MET FEI Tecnai G2-20 twin microscope equipped with LaB_6 source and providing an electron beam with an acceleration voltage of 200 kV. Samples were dry-deposed on a copper grid without any other treatment. Temperature-programmed reductions were operated on a Micromeritics AutoChem II 2920 chemisorption analyzer. A catalyst mass, fixed to ~40 mg, was inserted in a quartz reactor and degassed under inert gas. A flow of 5 vol.% H_2/N_2 was stabilized at a total flow rate of 50 mL·min^{-1}, and the catalyst was heated from 40 °C to 900 °C at a temperature increase rate of 10 °C·min^{-1} (K and P parameters respectively of 88 s and 15 °C). X-ray photoelectron spectroscopy analysis was carried out using a Kratos AXIS Ultra DLD apparatus with a monochromated Al Kα (1487 eV) source. Processing was performed using CasaXPS software, spectra being energy-corrected according to the main C 1s peak positioned at 284.8 eV.

3.3. Catalytic Activity Measurements

Catalytic performances were evaluated in the gas phase toluene oxidation reaction. A mass of catalyst equal to 0.200 g was positioned in a fixed bed reactor and was heated from 25 °C to 330 °C for30 min (2 °C·min^{-1}) under synthetic air flow. Then, the reaction flow, composed of 1000 ppmv of toluene diluted in synthetic air, was stabilized at a flow rate of 100 mL·min^{-1} (given a Weight Hourly Space Velocity of 30,000 mL·h^{-1}·g^{-1}) and the reactor temperature was allowed to decrease from 300 °C to 150 °C at a constant temperature decrease rate of −0.5 °C·min^{-1}. Stability experiments were performed during 70 h at a constant temperature of 285 °C. Exhaust gases were analyzed by gas chromatography and the results were expressed in terms of toluene conversion into carbon dioxide: $X(\%) = 100 \cdot [CO_2]_{out}/(7 \cdot [C_7H_8]_{in})$. To quantify the resistance against deactivation an activity coefficient a_{285} was defined as the ratio between the toluene conversion after 70 h reaction to that at initial time.

4. Conclusions

Mn- and Fe-containing perovskite were synthesized by a three-step reactive grinding process followed by calcination at 400 °C for 3 h. The reactive grinding process involved a first step resulting in the obtention of a crystalline perovskite at high temperature (solid state reaction) followed by a second step of HEBM affording a mean crystal size decrease to end up with a LBEM step performed in wet conditions for surface area improvement. Nanocrystalline $LaMnO_{3.15}$ and $LaFeO_3$ were obtained (crystal size range of 14–22 nm obtained after milling steps). The most efficient LEBM was found for the $LaFeO_3$ formulation allowing it to reach a SSA of 18.8 m^2·g^{-1}. It was noted that the TM reducibility was significantly affected by the HEBM step through the lowering of the crystal size leading to a decrease of the temperature of reduction of the Mn(+IV) and Fe(+III) species. XPS analysis evidenced the important effect of the grinding steps on the surface composition. Catalytic tests provided evidence of the importance of the catalyst SSA for the conversion of toluene. Then, while comparable activation energies were estimated for both perovskite compositions (Mn and Fe), from the overall results it was found that: (i) a higher intrinsic activity for the Mn-compositions than for the Fe-compositions due to a better redox behavior; (ii) a higher weight activity for Fe-compositions due to higher surface areas. Finally, while Mn-formulations deactivated significantly (−30% after 70 h on stream), far better stability was reported for Fe-formulations.

Author Contributions: B.H. prepared the materials, conducted the experiments and wrote the first draft of the paper. J.-F.L., S.R. and H.A. supervised the work. All authors contributed to the data interpretation, the discussion and the revision of the paper.

Funding: B.H. thanks Laval University (Canada) and Lille University (France) for funding his joint PhD. The "DepollutAir" project (grant number 1.1.18) of the European Program INTERREG V France-Wallonie-Flanders (FEDER), Chevreul institute (FR 2638), Ministère de l'Enseignement Supérieur et de la Recherche and Région Hauts-de-France are acknowledged for the funding and their support for this work.

Conflicts of Interest: The authors declare no conflict of interest.

Appendix A

Table A1. Binding energies of LaMnO$_{3.15}$ and LaFeO$_3$ samples' XPS components.

Sample	BE (eV)			
	La3d$_{5/2}$	Mn2p$_{3/2}$	Fe2p$_{3/2}$	O1s [1]
LaMn_SSR	834.7	641.6		529.3/531.2/533.0
LaMn_HEBM	834.1	641.6		529.2/531.3/533.3
LaMn_LEBM	833.9	641.7		529.4/531.3/533.5
LaFe_SSR	833.9		710.8	529.3/531.4/533.1
LaFe_HEBM	833.7		710.6	529.0/531.2/532.9
LaFe_LEBM	833.6		709.6	529.2/531.3/533.0

[1] values corresponding to O$_I$/O$_{II}$/O$_{III}$ type species.

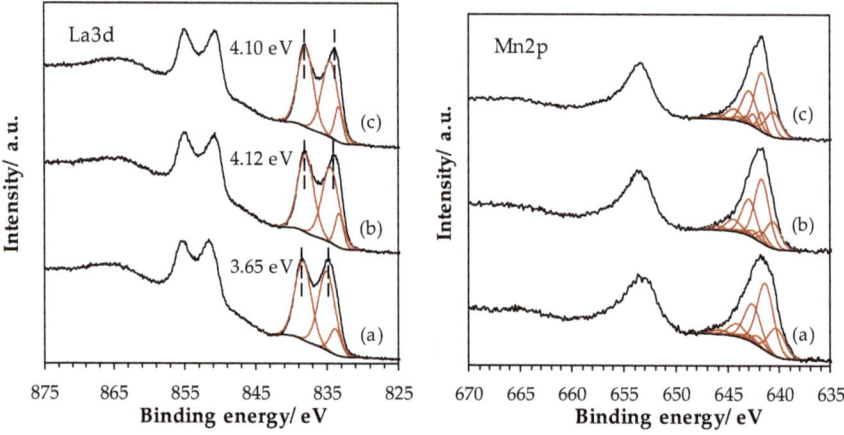

Figure A1. High-resolution spectra of La 3d and Mn 2p core level spectra for LaMnO$_{3.15}$ samples: (a) SSR, solid state reaction; (b) HEBM, high-energy ball milling; (c) LEBM, low-energy ball milling.

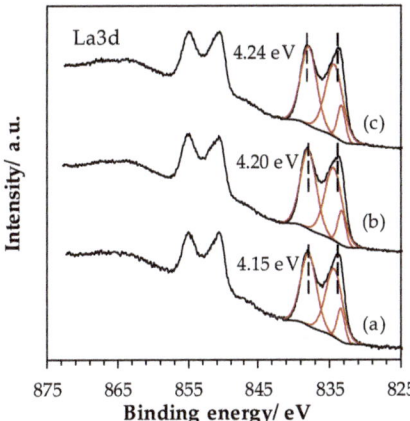

Figure A2. High-resolution spectra of La 3d core level spectra for LaFeO$_3$ samples: (a) SSR, solid state reaction; (b) HEBM, high-energy ball milling; (c) LEBM, low-energy ball milling.

References

1. He, C.; Cheng, J.; Zhang, X.; Douthwaite, M.; Pattisson, S.; Hao, Z. Recent Advances in the Catalytic Oxidation of Volatile Organic Compounds: A Review Based on Pollutant Sorts and Sources. *Chem. Rev.* **2019**, *119*, 4471–4568. [CrossRef] [PubMed]
2. Publications office of the European Union website. European Union Emission Inventory Report 1990–2015 under the UNECE Convention on Long-Range Transboundary Air Pollution (LRTAP). Available online: https://publications.europa.eu/s/kInt (accessed on 6 March 2019).
3. Donald, J.M.; Hooper, K.; Hopenhayn-Rich, C. Reproductive and developmental toxicity of toluene: A review. *Environ. Health Perspect.* **1991**, *94*, 237–244. [CrossRef] [PubMed]
4. Hamilton, J.F.; Webb, P.J.; Lewis, A.C.; Reviejo, M.M. Quantifying small molecules in secondary organic aerosol formed during the photo-oxidation of toluene with hydroxyl radicals. *Atmos. Environ.* **2005**, *39*, 7263–7275. [CrossRef]
5. Eur-LEX European Union Law website. Official Journal of the European Union, Commission Directive 2006/15/EC. Available online: https://eur-lex.europa.eu/eli/dir/2006/15/oj (accessed on 6 March 2019).
6. Amann, M.; Lutz, M. The revision of the air quality legislation in the European Union related to ground-level ozone. *J. Hazard. Mater.* **2000**, *78*, 41–62. [CrossRef]
7. Liotta, L.F. Catalytic oxidation of volatile organic compounds on supported noble metals. *Appl. Catal. B Environ.* **2010**, *100*, 403–412. [CrossRef]
8. Spivey, J.J.; Butt, J.B. Literature review: Deactivation of catalysts in the oxidation of volatile organic compounds. *Catal. Today* **1992**, *11*, 465–500. [CrossRef]
9. Ihm, S.-K.; Jun, Y.-D.; Kim, D.-C.; Jeong, K.-E. Low-temperature deactivation and oxidation state of Pd/γ-Al$_2$O$_3$ catalysts for total oxidation of n-hexane. *Catal. Today* **2004**, *93–95*, 149–154. [CrossRef]
10. Johnson Matthey website. PGM Market Report. Available online: http://www.platinum.matthey.com/ (accessed on 6 March 2019).
11. Royer, S.; Duprez, D. Catalytic Oxidation of Carbon Monoxide over Transition Metal Oxides. *ChemCatChem* **2011**, *3*, 24–65. [CrossRef]
12. Quiroz Torres, J.; Royer, S.; Bellat, J.-P.; Giraudon, J.-M.; Lamonier, J.-F. Formaldehyde: Catalytic Oxidation as a Promising Soft Way of Elimination. *ChemSusChem* **2013**, *6*, 578–592. [CrossRef]
13. Twu, J.; Gallagher, P.K. Chapter 1: Preparation of Bulk and Supported Perovskites. In *Properties and Applications of Perovskite-Type Oxides*, 1st ed.; Tejuca, L.G., Fierro, J.L.G., Eds.; CRC Press: Boca Raton, FL, USA, 1992; pp. 1–23.
14. Peña, M.A.; Fierro, J.L.G. Chemical Structures and Performance of Perovskite Oxides. *Chem. Rev.* **2001**, *101*, 1981–2018. [CrossRef]

15. Royer, S.; Duprez, D.; Can, F.; Courtois, X.; Batiot-Dupeyrat, C.; Laassiri, S.; Alamdari, H. Perovskites as substitutes of noble metals for heterogeneous catalysis: Dream or reality. *Chem. Rev.* **2014**, *114*, 10292–10368. [CrossRef] [PubMed]
16. Wachowski, L. Influence of the method of preparation on the porous structure of perovskite oxides. *Surf. Coat. Technol.* **1986**, *29*, 303–311. [CrossRef]
17. Royer, S.; Bérubé, F.; Kaliaguine, S. Effect of the synthesis conditions on the redox and catalytic properties in oxidation reactions of $LaCo_{1-x}Fe_xO_3$. *Appl. Catal. A Gen.* **2005**, *282*, 273–284. [CrossRef]
18. Zhang, C.; Guo, Y.; Guo, Y.; Lu, G.; Boreave, A.; Retailleau, L.; Baylet, A.; Giroir-Fendler, A. $LaMnO_3$ perovskite oxides prepared by different methods for catalytic oxidation of toluene. *Appl. Catal. B Environ.* **2014**, *148–149*, 490–498. [CrossRef]
19. Baláž, P.; Achimovičová, M.; Baláž, M.; Billik, P.; Cherkezova-Zheleva, Z.; Manuel Criado, J.; Delogu, F.; Dutková, E.; Gaffet, E.; José Gotor, F.; et al. Hallmarks of mechanochemistry: From nanoparticles to technology. *Chem. Soc. Rev.* **2013**, *42*, 7571–7637. [CrossRef] [PubMed]
20. Alamdari, H.; Royer, S. Chapter 2: Mechanochemistry. In *Perovskites and Related Mixed Oxides*; Granger, P., Parvulescu, V.I., Prellier, W., Eds.; Wiley-VCH Verlag GmbH & Co. KGaA: Weinheim, Germany, 2015; pp. 25–46.
21. Ciotonea, C.; Averlant, R.; Rochard, G.; Mamede, A.S.; Giraudon, J.M.; Alamdari, H.; Lamonier, J.-F.; Royer, S. A Simple and Green Procedure to Prepare Efficient Manganese Oxide Nanopowder for the Low Temperature Removal of Formaldehyde. *ChemCatChem* **2017**, *9*, 2366–2376. [CrossRef]
22. Laassiri, S.; Bion, N.; Duprez, D.; Royer, S.; Alamdari, H. Clear microstructure–performance relationships in Mn-containing perovskite and hexaaluminate compounds prepared by activated reactive synthesis. *Phys. Chem. Chem. Phys.* **2014**, *16*, 4050–4060. [CrossRef] [PubMed]
23. Levasseur, B.; Kaliaguine, S. Methanol oxidation on $LaBO_3$ (B = Co, Mn, Fe) perovskite-type catalysts prepared by reactive grinding. *Appl. Catal. A Gen.* **2008**, *343*, 29–38. [CrossRef]
24. Kaliaguine, S.; Van Neste, A. Process for Synthesizing Metal Oxides and Metal Oxides Having A Perovskite or Perovskite-Like Crystal Structure. U.S. Patent 6,770,256, 3 August 2004.
25. Ghasdi, M.; Alamdari, H.; Royer, S.; Adnot, A. Electrical and CO gas sensing properties of nanostructured $La_{1-x}Ce_xCoO_3$ perovskite prepared by activated reactive synthesis. *Sens. Actuators B Chem.* **2011**, *156*, 147–155. [CrossRef]
26. Thommes, M.; Kaneko, K.; Neimark, A.V.; Olivier, J.P.; Rodriguez-Reinoso, F.; Rouquerol, J.; Sing, K.S. Physisorption of gases, with special reference to the evaluation of surface area and pore size distribution. *Pure Appl. Chem.* **2015**, *87*, 1051–1069. [CrossRef]
27. Laassiri, S.; Duprez, D.; Royer, S.; Alamdari, H. Solvent free synthesis of nanocrystalline hexaaluminate-type mixed oxides with high specific surface areas for CO oxidation reaction. *Catal. Sci. Technol.* **2011**, *1*, 1124–1127. [CrossRef]
28. Kaliaguine, S.; Van Neste, A.; Szabo, V.; Gallot, J.E.; Bassir, M.; Muzychuk, R. Perovskite-type oxides synthesized by reactive grinding: Part I. Preparation and characterization. *Appl. Catal. A.* **2001**, *209*, 345–358. [CrossRef]
29. Vogel, E.M.; Johnson, D.W., Jr.; Gallagher, P.K. Oxygen Stoichiometry in $LaMn_{1-x}Cu_xO_{3+y}$ by Thermogravimetry. *J. Am. Ceram. Soc.* **1977**, *60*, 31–33. [CrossRef]
30. Irusta, S.; Pina, M.P.; Menéndez, M.; Santamaría, J. Catalytic Combustion of Volatile Organic Compounds over La-Based Perovskites. *J. Catal.* **1998**, *179*, 400–412. [CrossRef]
31. Faye, J.; Baylet, A.; Trentesaux, M.; Royer, S.; Dumeignil, F.; Duprez, D.; Valange, S.; Tatibouët, J.M. Influence of lanthanum stoichiometry in La1-xFeO3-δ perovskites on their structure and catalytic performance in CH_4 total oxidation. *Appl. Catal. B Environ.* **2012**, *126*, 134–143. [CrossRef]
32. Zhang, C.; Wang, C.; Zhan, W.; Guo, Y.; Guo, Y.; Lu, G.; Baylet, A.; Giroir-Fendler, A. Catalytic oxidation of vinyl chloride emission over $LaMnO_3$ and $LaB_{0.2}Mn_{0.8}O_3$ (B = Co, Ni, Fe) catalysts. *Appl. Catal. B Environ.* **2013**, *129*, 509–516. [CrossRef]
33. Fierro, J.L.G.; Gonzalez Tejuca, L. Non-stoichiometric surface behaviour of LaMO3 oxides as evidenced by XPS. *Appl. Surf. Sci.* **1987**, *27*, 453–457. [CrossRef]
34. Deng, J.; Zhang, L.; Dai, H.; He, H.; Au, C.T. Strontium-Doped Lanthanum Cobaltite and Manganite: Highly Active Catalysts for Toluene Complete Oxidation. *Ind. Eng. Chem. Res.* **2008**, *47*, 8175–8183. [CrossRef]

35. Fierro, J.L.G. Structure and composition of perovskite surface in relation to adsorption and catalytic properties. *Catal. Today* **1990**, *8*, 153–174. [CrossRef]
36. Sunding, M.F.; Hadidi, K.; Diplas, S.; Løvvik, O.M.; Norby, T.E.; Gunnæs, A.E. XPS characterisation of in situ treated lanthanum oxide and hydroxide using tailored charge referencing and peak fitting procedures. *J. Electron. Spectros. Relat. Phenom.* **2011**, *184*, 399–409. [CrossRef]
37. Biesinger, M.C.; Payne, B.P.; Grosvenor, A.P.; Lau, L.W.M.; Gerson, A.R.; Smart, R.S.C. Resolving surface chemical states in XPS analysis of first row transition metals, oxides and hydroxides: Cr, Mn, Fe, Co and Ni. *Appl. Surf. Sci.* **2011**, *257*, 2717–2730. [CrossRef]
38. Bagus, P.S.; Broer, R.; Ilton, E.S. A new near degeneracy effect for photoemission in transition metals. *Chem. Phys. Lett.* **2004**, *94*, 150–154. [CrossRef]
39. Galakhov, V.R.; Demeter, M.; Bartkowski, S.; Neumann, M.; Ovechkina, N.A.; Kurmaev, E.Z.; Lobachevskaya, N.I.; Mukovskii, Y.M.; Mitchell, J.; Ederer, D.L. Mn exchange splitting in mixed-valence manganites. *Phys. Rev. B* **2002**, *65*, 113102. [CrossRef]
40. McCarty, J.G.; Wise, H. Perovskite catalysts for methane combustion. *Catal. Today* **1990**, *8*, 231–248. [CrossRef]
41. Pecchi, G.; Jiliberto, M.G.; Delgado, E.J.; Cadús, L.E.; Fierro, J.L.G. Effect of B-site cation on the catalytic activity of La1−xCaxBO3 (B = Fe, Ni) perovskite-type oxides for toluene combustion. *J. Chem. Technol. Biotechnol.* **2011**, *86*, 1067–1073. [CrossRef]
42. Behar, S.; Gómez-Mendoza, N.A.; Gómez-García, M.Á.; Świerczyński, D.; Quignard, F.; Tanchoux, N. Study and modelling of kinetics of the oxidation of VOC catalyzed by nanosized Cu–Mn spinels prepared via an alginate route. *Appl. Catal. A Gen.* **2015**, *504*, 203–210. [CrossRef]

© 2019 by the authors. Licensee MDPI, Basel, Switzerland. This article is an open access article distributed under the terms and conditions of the Creative Commons Attribution (CC BY) license (http://creativecommons.org/licenses/by/4.0/).

MDPI
St. Alban-Anlage 66
4052 Basel
Switzerland
Tel. +41 61 683 77 34
Fax +41 61 302 89 18
www.mdpi.com

Catalysts Editorial Office
E-mail: catalysts@mdpi.com
www.mdpi.com/journal/catalysts